Introduction to t..
Calculus of Variations

U. ... n-Manderscheid

Translated ... P.G. Er... rom

CHAPMAN & HALL

London · New York · Tokyo · Melbourne · Madras

UK	Chapman & Hall, 2–6 Boundary Row, London SE1 8HN
USA	Chapman & Hall, 29 West 35th Street, New York NY10001
JAPAN	Chapman & Hall Japan, Thomson Publishing Japan, Hirakawacho Nemoto Building, 7F, 1–7–11 Hirakawa-cho, Chiyoda-ku, Tokyo 102
AUSTRALIA	Chapman & Hall Australia, Thomas Nelson Australia, 102 Dodds Street, South Melbourne, Victoria 3205
INDIA	Chapman & Hall India, R. Seshadri, 32 Second Main Road, CIT East, Madras 600 035

Original German language edition Ursula Brechtken-Manderscheid,
Einführung in die Variationsrechnung

English edition 1991

Printed and Bound in Great Britain by
T. J. Press (Padstow) Ltd, Padstow, Cornwall

ISBN 0 412 36690 8 (HB)
0 412 36700 9 (PB)

British Library Cataloguing in Publication Data

Brechtken-Manderscheid, Ursula
Introduction to the calculus of variations.
I. Title
515

ISBN 0-412-36690-8
ISBN 0-412-36700-9 pbk

Library of Congress Cataloging-in-Publication Data

Brechtken-Manderscheid, U.
[Einführung in die Variationsrechnung. English]
Introduction to the calculus of variations / U. Brechtken
-Manderscheid : translated by P.G. Engstrom.
p. cm. — (Chapman & Hall mathematics)
Translation of: Einführung in die Variationsrechnung.
Includes bibliographical references and index.
ISBN 0-412-36690-8. — ISBN 0-412-36700-9 (pbk.)
1. Calculus of variations. I. Title. II. Series.
QA315.B8313 1991
515'.64—dc20 91-12280
CIP

£30 DAW

Contents

Preface

The objects of the calculus of variations are optimization problems which may be characterized as follows: Given a set of functions, curves or surfaces, find those which possess a certain maximal property. One can, for example, consider all closed plane curves k of length 1 meter and ask which of these curves is the boundary for the largest area. Another variational problem is this: What form will be taken by a heavy chain which is supported at its ends? If the centre of mass of the chain is as low as possible, then the chain has the minimal property. Similar problems appear in mathematics, in physics, in technology and in economics. Just as with problems in extrema for ordinary functions, so in the calculus of variations one works principally with necessary and sufficient conditions.

Variational problems were formulated even in antiquity. But only after the differential and integral calculus had been developed by Leitnitz (1667–1748) and Newton (1642–1727) could a systematic investigation begin. After the Bernoulli brothers, John (1667–1748) and James (1654–1705), had popularized the problems (cf. [1]), Euler (1707–1783) derived the fundamental necessary conditions and thereby founded the new discipline. The youthful Lagrange inherited the task of unifying the various problems. After him the use of the calculus of variations receded. Euler had coined the name 'calculus of variations'.

The strong interest in the young science can be explained by the world view of the time, which found its expression in various philosophically and theologically founded extremal principles, like the least action principle of Maupertuis (1698–1759). Men were of the opinion that 'nature rules all according to the most beautiful and best, and if one proceeds from this conception, then one can intellectually grasp all of the individual phenomena of the exact sciences' (from Plato's *Phaidon*, cited in [6]). We will not go further into the history of the calculus of variations but only mention some additional researchers who are especially important for the discipline. They are Legendre (1752–1833), Jacobi (1804–1851), Weierstrass (1815–1897) and Hilbert (1862–1943).

In order to make concrete, useful statements it is necessary to divide variational problems into classes for which a single method of solution is possible. The same idea, however, is at the root of the necessary and sufficient conditions for the different types of variational problems. Thus, we will develop these conditions first for the simplest type of variational problem; that is, we will describe thoroughly all necessary and sufficient conditions for this problem. In the second half of this book the already developed methods will be carried over to the most important other classes. Each of the Chapters 2 through 6 treat, therefore, one simple condition for the simplest problem, and each of the Chapters 7 through 11 treat a special type of variational problem. The final Chapter 12 introduces newer methods of approximation of solutions of extremal problems.

Professor W. Velte has looked through the manuscript and has given me many valuable suggestions. I would like, therefore, to thank him here. My thanks also go to Mrs. H. Grieco who with great patience and care has assembled the text of this book.

U. Brechten-Manderscheid
Würzburg, 1983

Notation

Numbers in square brackets refer to the corresponding numbered references in the Bibliography. The closed interval $\{x \in \mathbf{R} : a \leq x \leq b\}$ is denoted by $[a, b]$ and the open interval $\{x \in \mathbf{R} : a < x < b\}$ by (a, b). The symbol $\mathbf{R}$ denotes the real numbers.

Functions will generally be denoted by the function symbol. So, for example, the function $y : [a, b] \mapsto \mathbf{R}$ will be denoted by y rather than $y(x)$. A vector function is recognized by the bar under the letter designation. So the vector function $\underline{y} : [a, b] \mapsto \mathbf{R}^n$ is written $\underline{y}(x) = (y_1(x), y_2(x), \ldots, y_n(x)) = (y_i(x))$, where the y_i, $i = 1, 2, 3, \ldots, n$, are the component functions.

The partial derivative with respect to x of a function f of several variables is denoted by $\frac{\partial}{\partial x} f$ or, more concisely, f_x. Similar expressions are used for second and higher partial derivatives. Thus, the second mixed partial derivative of y with respect to x and α will be denoted by $\frac{\partial^2 y}{\partial x \partial \alpha}$ or $y_{x\alpha}$.

In the calculus of variations we frequently have to deal with compositions of functions. We may have, for example, $f : R^3 \mapsto \mathbf{R}$, where f has continuous partial derivatives with respect to the three variables x, y and y', and $y_0 : [a, b] \mapsto \mathbf{R}$ has a continuous second derivative with respect to x. (As usual, the derivative of y_0 is denoted by y_0'.) Then the composite function $f(\cdot, y_0, y_0')$ can be formed: $x \mapsto f(x, y_0(x), y_0'(x))$. The derivative of this composite function at the point x is written $\frac{d}{dx} f(\cdot, y_0, y_0')(x)$.

According to the chain rule, this derivative is equal to

$$\begin{aligned} f_x(x, y_0(x), y_0'(x)) + f_y(x, y_0(x), y_0'(x))y_0' \\ + f_{y'}(x, y_0(x), y_0'(x))y_0''. \end{aligned}$$

One must not, of course, confuse $\frac{d}{dx}f(\cdot, y_0, y_0')$ with the partial derivative of f with respect to x.

1

Introduction to the problem

Variational problems confront us in many areas in daily life as well as in the natural sciences, economics and technology. The variety of problems is so great that it is not sensible to describe or to treat all variational problems in one single way. We will first introduce the most important class of problems through some classical examples. In Section 1.2 the basic concepts will be set down. Then in Chapters 2 through 6 we will derive methods of solution for this type of problem, though not before we have in Section 1.3 considered whether variational problems always have a solution. If one knows in advance that a given problem is solvable, then the solution can often be found quickly. Yet, there are simple problems which possess no solution.

1.1 EXAMPLES

(1.1) The problem of the shortest curve: On a surface, for example, the surface of the earth, find the shortest path joining given points A and B. (1.2) The smallest area of a given surface of revolution: Two points A and B and a line are given. Let A and B be joined by a curve k. How shall k be chosen so that the area of the surface of revolution obtained by revolving k about the given line is minimized (cf. Fig. 1.1)?

The following problem was proposed by John Bernoulli in 1696. With it the systematic investigation of variational prob-

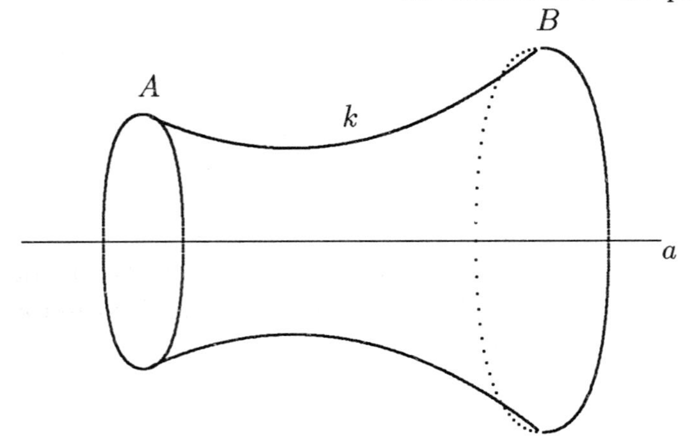

Fig. 1.1. Minimal surface of revolution.

lems began.

(1.3) The Bernoulli problem: Suppose that A and B are given points in a vertical plane and let M be a movable point on the path AMB. The path shall be chosen in such a way that, if M moves along AMB from A to B under the influence of its own weight, then the time of travel will be smallest.

The next problem is much older. It appears in the saga of the founding of the city of Carthage and has to do with the minimizing of the area of a piece of land.

(1.4) The problem of Dido: Among all closed curves of length L lying in the plane, which one will enclose the largest area?

Now we proceed to formulate these problems mathematically; we shall see which concepts need to be developed.

The problem (1.1) *of the shortest curve joining two points:* Let M be the set of all paths on a given surface F and joining the points A and B. Designate the length of a path w with $L(w)$. Thus we seek a path $w_0 \in M$ having the smallest length

$$L(w_0) = \min_{w \in M} L(w).$$

We deal here clearly with an extremal problem.

We do not want to pursue the most general case (cf. Chapter 9), but rather assume that F is a plane in which the points A and B, with respect to a rectangular coordinate system, have coordinates $A = (a, y_a)$, $B = (b, y_b)$; $a < b$.

We now require that each path w shall be representable by a smooth function y of x (we say a function is *smooth* if its first derivative is continuous; it is *n-fold smooth* if its nth derivative is continuous). Then the length of w is given by the formula

$$L(w) = L(y) = \int_a^b \sqrt{1 + y'^2(x)}\, dx. \tag{1.5}$$

The value of L depends on the function y so that L represents a 'function of a function'. We call such functions *functionals.*

Problem (1.1) may now be stated as follows: Among all smooth functions y with the boundary conditions $y(a) = y_a$, $y(b) = y_b$, find that one for which the integral has the smallest value.

The problem (1.2) *of the surface of revolution having the smallest area:* We choose a rectangular coordinate system so that the plane is the xy-plane. Let the x-axis be the axis of rotation and let $A = (a_1, a_2)$, $B = (b_1, b_2)$, $a_1 < b_1$; $y : [a_1, b_1] \mapsto \mathbb{R}$ shall be a smooth function satisfying the boundary conditions $y(a_1) = a_2$, $y(b_1) = b_2$. Now y describes a curve k which connects A and B. If k is revolved about the x-axis, there will be generated an area given by

$$J(y) = 2\pi \int_{a_1}^{b_1} |\, y(x)\, | \sqrt{1 + y'^2(x)}\, dx. \tag{1.6}$$

Consider for a moment the function y_0 for which the quantity $J(y_0)$, the area of the surface of revolution, is smallest. Someone may raise the objection that there are curves from A to B which are not representable by a function (cf. Fig. 1.2). There are also curves which have corners and which, even though represented by a function, are not represented by a smooth function. In any event, the problem formulated above is an interesting

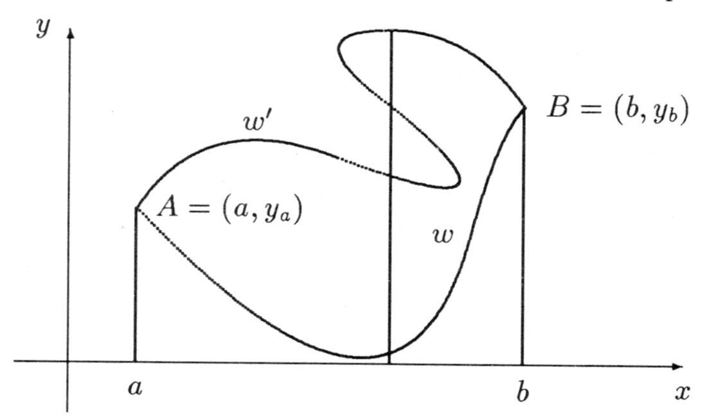

Fig. 1.2. The path w can be represented by a function y; the path w', on the contrary, cannot.

one, partly because we can convince ourselves that the excluded curves never come into question as possible solutions.

The Bernoulli problem (1.3): We choose a rectangular coordinate system in which the positive y-axis is directed in the direction of the gravitational force (cf. Fig. 1.3). Again set $A = (a, y_a)$ and $B = (b, y_b)$. We make use here of the law of the conservation of energy. For all times t along the path, the sum of the kinetic and potential energy is constant:

$$\tfrac{1}{2}mv^2(t) + mgh(t) = \tfrac{1}{2}mv^2(t_0) - mgy_a = c = \text{constant.} \quad (1.7)$$

Here g is the gravitational constant, $v(t)$ is the velocity and $-h(t)$ is the y-coordinate distance. Thus, h is the height of the moving point-mass at the time t. The calculations become significantly simpler if we restrict the paths to those representable by smooth functions y of x, $y : [a, b] \mapsto \mathbb{R}$, which also have the property that to each value x there is at most one point on the path having the abcissa x. We denote the length of the path between t_0 and t with $s(t)$. Then

$$v(t) = \frac{ds(t)}{dt} = \sqrt{\frac{2c}{m} - 2gh(t)}\,.$$

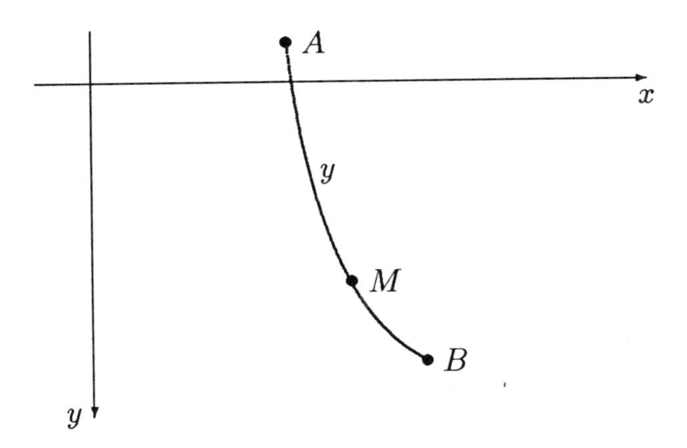

Fig. 1.3. Path of least time of descent.

The time required for the point-mass to travel from A to B is

$$\begin{aligned} T = T(y) &= \int_{t_A}^{t_B} dt \\ &= \int_{s(t_A)}^{s(t_B)} \frac{ds}{v(t)} \qquad (1.8) \\ &= \int_a^b \frac{\sqrt{1 + y'^2(x)}}{\sqrt{\frac{2c}{m} + 2gy(x)}}\, dx. \end{aligned}$$

Thus we seek a function y with boundary values $y(a) = y_a$ and $y(b) = y_b$ so that $T(y)$ is smallest.

The problem of Dido (1.4): First, observe that only convex curves need be considered here (a plane closed curve is said to be *convex* if the line segment joining any two points on the curve lies entirely within the area bounded by the curve (cf. Figs. 1.4 and 1.5)). To show this we choose a rectangular coordinate system so that the x-axis divides the area G within the

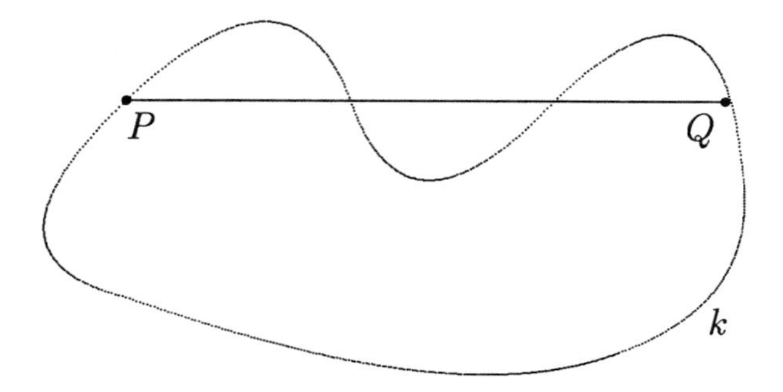

Fig. 1.4. This curve is *not* convex. The curve in Fig. 1.5 is.

curve k into two equal areas G^+ and G^- (cf. Fig. 1.5). If the circumference of G^+ is not precisely the same as that of G^-, then the curve cannot contain the largest possible area. For if the region having the smaller circumference is reflected in the x-axis, then we obtain a curve $\tilde{k}$ enclosing the same area but having a smaller total length. If the length of $\tilde{k}$ is now increased so that its length is L, then the new area enclosed by this curve will be larger than that enclosed by the curve k. We need, therefore, only investigate those curves k for which G^+ and G^- have the same circumference. That part k^+ of the curve k which belongs to the boundary of G^+ has the length $L/2$. We introduce the parameter s, where s represents the length of the curve

$$k^+ : s \in \left[0, \tfrac{L}{2}\right] \mapsto (x(s), y(s)),$$
$$y(0) = y\left(\tfrac{L}{2}\right) = 0, \quad y(s) \geq 0 \text{ for } s \in \left[0, \tfrac{L}{2}\right].$$

The surface area $I(G^+)$ of G^+ is, therefore,

$$I(G^+) = \int_0^{\frac{L}{2}} y(s)\sqrt{1 - y'^2(s)}\, ds. \tag{1.9}$$

We want y so that $I(G^+)$ is as large as possible.

Note: There is no less generality if in the following pages we consider only minimum problems, since the maximum of a function F is only the minimum of $-F$.

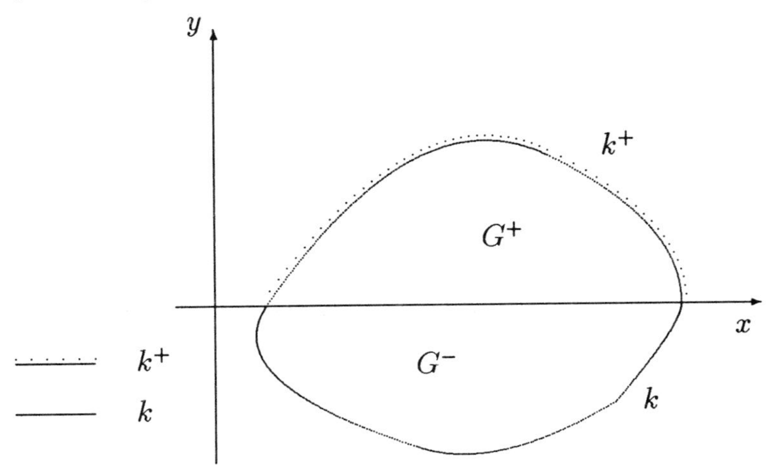

Fig. 1.5. A convex figure.

1.2 DEFINITION OF THE MOST IMPORTANT CONCEPTS

It is typical of the examples that the functional whose value is to be minimized or maximized appears in the form of an integral with given upper and lower limits:

$$J(y) = \int_a^b f(x, y(x), y'(x))\, dx. \tag{1.10}$$

This integral can certainly be evaluated for many possible functions y. But for extremal problems only certain functions are of interest. These we will call *admissible functions*. Thus, each function which is defined and smooth on $[a, b]$ and which takes on the values y_a and y_b at the endpoints a and b will be admissible. In Section 2.4 we will consider further the set of admissible functions and give attention to various related questions.

In order to set down the particulars of a variational problem, we use the following notation:

$$J(y) = \int_a^b f(x, y(x), y'(x))\, dx \to \min.$$

Boundary conditions: $y(a) = y_a$, $y(b) = y_b$.

This is the simplest and most important type of variational problem. We will develop this type in Chapters 2 through 6 and learn to understand the methods of solution belonging to it. Thereafter, these methods will be generalized to other kinds of variational problems.

The integrand f with arguments x, y and y' will be assumed to satisfy the following conditions: The function f shall be defined for all points (x, y) in an open set G in the xy-plane and for all $y' \in \mathbb{R}$. Furthermore, f shall have continuous partial derivatives up to and including the third order with respect to each variable. Generally, G will be in $\mathbb{R}^2$ or will be defined by an expression of the form

$$\{(x, y) : a_1 < x < b_1,\ \varphi_1(x) < y < \varphi_2(x)\},$$

where $a_1 < a$, $b_1 > b$ and $\varphi_1 < \varphi_2$ are continuous functions on (a_1, b_1).

If G is not all of $\mathbb{R}^2$, then only those functions y for which the variational integral is determined are admissible; i.e. those for which $(x, y(x)) \in G$ for all $x \in [a, b]$.

The variational problem may now be posed as follows: Among all admissible functions y, determine that one for which the variational integral J takes on its minimal value. We call an admissible function y_0 a *solution* of the variational problem or an *absolute minimum* of J if for all admissible functions we have $J(y_0) \leq J(y)$. In the event that for all admissible functions $y \neq y_0$ we have $J(y_0) < J(y)$, then y_0 is called a *proper absolute minimum* of J.

Just as with the extreme value problems of elementary analysis, so here we look not only for absolute minima but also for *relative minima*. But what does 'relative' mean in this connection? To answer this we must clarify what it means for two admissible functions to lie sufficiently close to each other.

There are different possibilities for defining the "distance" between two functions y_1 and y_2. For our applications the most important are the distances d_0 and d_1. The *distance* d_0 is defined

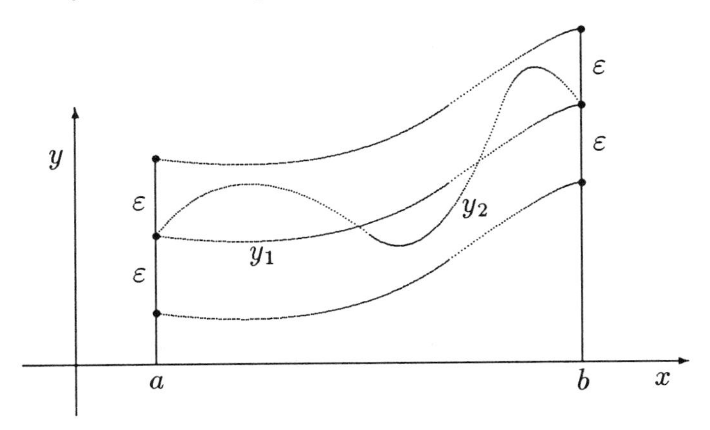

Fig. 1.6. The ε-band for y_1.

for continuous functions y_1 and y_2 on (a, b) by

$$d_0(y_1, y_2) = \max_{x \in [a,b]} \mid y_1(x) - y_2(x) \mid . \tag{1.11}$$

For smooth functions y_1 and y_2 on the interval $[a, b]$ we define the *distance* d_1 by

$$d_1(y_1, y_2) = \sup_{x \in [a,b]} (\mid y_1(x) - y_2(x) \mid + \mid y_1{}'(x) - y_2{}'(x) \mid). \tag{1.12}$$

We can illustrate in the following way which admissible functions y_2 in the xy-plane have a d_0-distance from y_1 which is less than ε. For all $x \in [a, b]$ we must have $-\varepsilon < y_2 - y_1 < \varepsilon$; i.e. the graph of y_2 must lie within an ε-band S_ε about the graph of y_1.

According to the definition, the d_1-distance between two functions is never less than their d_0-distance. Figure 1.6 shows that among the functions y with the property $d_0(y, y_1) < \varepsilon$ there are functions y whose d_1-distance from y_1 is arbitrarily large. This follows because the difference between the derivatives of y and y_1 may differ sharply. We designate an admissible function y_0 to be a (*relative*) *strong solution* of the variational problem if there is an $\varepsilon > 0$ such that

$$J(y_0) \leq J(y)$$

for all admissible functions y for which $d_0(y_0, y) < \varepsilon$.

Analogously, an admissible function is said to be a (*relative*) *weak solution* of the variational problem if there is an $\varepsilon > 0$ such that

$$J(y_0) \leq J(y)$$

for all admissible functions y with $d_1(y_0, \dot{y}) < \varepsilon$.

A strong solution of a variational problem is also a weak solution of the problem.

1.3 THE QUESTION OF THE EXISTENCE OF SOLUTIONS

Before we derive the necessary and sufficient conditions for solutions of variational problems, we make the observation that not every variational problem possesses a solution. For functions of one or more variables the same difficulty exists. Thus, for example, the function $f : (0,1) \mapsto \mathbb{R}$, $f(x) = x$, has no minimum and no maximum on the open interval $(0,1)$. On the other hand, the following theorem holds.

(1.13) Weierstrass Theorem
Every continuous function f which is defined and continuous on a closed and bounded domain $M \subset \mathbb{R}^n$ has a minimum and a maximum.

It depends, therefore, on the domain of definition of the function. For variational problems, however, the assumptions of a generalized Weierstrass theorem with respect to the domain of definition are for the most part not fulfilled. One can, of course, alter the problem so that these assumptions are fulfilled; i.e. one can extend the functional J to a more appropriate domain of definition. By doing so, however, the original problem is often changed in a non-trivial way. For the following two problems there is no solution.

(1.14) Example

Among all smooth functions y which are defined on $[-1,1]$ and which satisfy the boundary conditions $y(-1) = -1$ and $y(1) = 1$, find the function(s) for which the integral

$$J(y) = \int_{-1}^{1} x^4 y'^2(x)\, dx$$

has the smallest value. Since the integrand cannot be negative, $J(y) \geq 0$ for all admissible functions y. One can easily check that the functions $y_n : [-1,1] \mapsto \mathbb{R}$, defined by

$$y_n(x) = \begin{cases} -1 & \text{if} \quad -1 \leq x < -\alpha_n, \\ -1 + n^3(x+\alpha_n)^2 & \text{if} \quad -\alpha_n \leq x < -\beta_n, \\ nx & \text{if} \quad -\beta_n \leq x < \beta_n, \\ 1 - n^3(x-\alpha_n)^2 & \text{if} \quad \beta_n \leq x < \alpha_n, \\ 1 & \text{if} \quad \alpha_n \leq x \leq 1, \end{cases},$$

$$\alpha_n = \frac{1}{n} + \frac{1}{4n^2} \quad \text{and} \quad \beta_n = \frac{1}{n} - \frac{1}{4n^2},$$

are continuous and differentiable. It is also clear that $\lim_{n\to\infty} J(y_n) = 0$. Thus, $\inf J(y) = 0$.

If, for a smooth function $y_0 : [-1,1] \mapsto \mathbb{R}$, the functional J has the value 0, then the integrand $x^4 {y_0'}^2(x)$ must be zero for all $x \in [-1,1]$; i.e. $y_0'(x)$ is identically zero on the interval and the function y_0 is a constant. But if it is a constant, it cannot fulfil the prescribed boundary conditions. There is, therefore, no admissible function y_0 such that $J(y_0) = 0$; the variational problem has no solution.

The following example shows that these difficulties arise not only from theoretical examples but also in examples from daily life.

(1.15) Example

A sailboat is to travel on a river (cf. Fig. 1.7) from point A to point B in the least possible time.

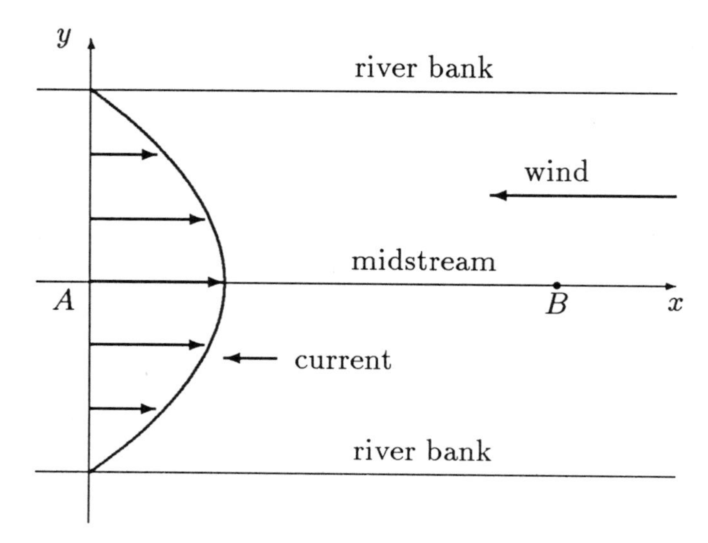

Fig. 1.7. The sailboat problem.

The current of the river is strongest in the middle of the stream. The wind, however, blows from B towards A and so opposes the motion of the sailboat. By investigating the different forces which act upon the boat, we see that the time T needed to travel from A to B is given by the formula

$$T = J(y) = \int_0^b \frac{1 + y'^2(x)}{(v_m + c)\sqrt{1 + y'^2(x)} - c + g(y(x))} \, dx, \quad (1.16)$$

where $y(x)$ is the deviation of the boat from the middle of the river when the boat is x km from A, v_m and c are constants, and $g(y)$ is the influence of the current.

Consider now the case where the flow of the river is zero. Here there is clearly an optimal strategy: The sailor tacks into the wind and within a band whose width is determined by the banks of the river; i.e. he maintains a prescribed, fixed angle with respect to the wind. If the flow of the river is not negligible, then the smaller the band along the middle of the river in which the boat tacks, the greater the advantage gained from the river's flow. There is no optimal course here, however, for in the limiting case the boat is restricted to travel along the line in the middle of the river, and here it cannot tack.

2

The Euler differential equation

In this chapter we recall the extreme value problem for a differentiable function Φ of one variable. A weak solution of the variational problem corresponds to a relative minimum for Φ and conversely. A differentiable function Φ can, however, have a relative minimum only if certain necessary conditions are satisfied. Since these conditions are basic for our theory, we recall their derivation for the reader.

If Φ at the point x_0 has a relative minimum, then it holds that

$$\Phi(x) - \Phi(x_0) \geq 0 \tag{2.1}$$

for all x in the domain of definition of Φ and sufficiently close to x_0. For these x values

$$\frac{\Phi(x) - \Phi(x_0)}{x - x_0} \begin{cases} \leq 0 & \text{if} \quad x - x_0 \leq 0, \\ \geq 0 & \text{if} \quad x - x_0 \geq 0. \end{cases}$$

If Φ is differentiable, then the difference quotient converges to $\Phi'(x_0)$ as $x \to x_0$. If x_0 is an interior point of the domain of definition, then

$$\lim_{x \to x_0^-} \frac{\Phi(x) - \Phi(x_0)}{x - x_0} = \Phi'(x_0) \leq 0$$

and

$$\lim_{x \to x_0^+} \frac{\Phi(x) - \Phi(x_0)}{x - x_0} = \Phi'(x_0) \geq 0.$$

Therefore, $\Phi'(x_0) = 0$. If, contrarily, x_0 is a boundary point of the domain of definition of Φ, then only one of the limits is defined and we obtain an inequality for $\Phi'(x_0)$. For example, if the right boundary point x_0 of the domain provides a relative minimum, then we must have $\Phi'(x_0) \leq 0$.

2.1 DERIVATION OF THE EULER DIFFERENTIAL EQUATION

We turn now to the first simple type of variational problem:

$$J(y) = \int_a^b f(x, y(x), y'(x))\, dx \to \min .$$

The admissible functions y are the smooth functions of $x \in [a, b]$ which have the prescribed boundary values y_a and y_b at the endpoints a and b. We show in this section that each solution of this variational problem must satisfy a certain differential equation, i.e. the Euler differential equation. This is an important piece of information which helps us determine the solutions of the variational problem. It means that we can find solutions of the variational problem only among the solutions of the Euler equation.

In order to derive this differential equation we assume that a solution y_0 of the variational problem has continuous first, second and third derivatives. The set of all such admissible functions is too large (more precisely said, the set is of infinite dimension). In order to deal effectively with the condition $J(y) - J(y_0) \geq 0$, we need simpler methods. We employ, therefore, the following device, due to Lagrange: We select from this larger set a subset of one-parameter functions and thereby relate the variational problem to an extreme value problem of a function of a single variable. This subset consists of the functions y_α, defined for $x \in [a, b]$ by

$$y_\alpha(x) = y_0(x) + \alpha Y(x), \quad |\, \alpha \,| < \alpha_0,$$

where α_0 is a positive, sufficiently small number and Y is a smooth function on $[a, b]$ which satisfies the boundary conditions

$$Y(a) = Y(b) = 0. \tag{2.2}$$

It should be made very clear that the functions y_α are admissible functions. Furthermore, each arbitrary function Y which is defined on $[a, b]$, which is smooth and which satisfies the boundary conditions, (2.2) leads to one of the above mentioned (one-parameter) subsets. The quantity αY, occasionally written δY, is called the *variation* (since y_0 is varied by the addition of αY). Hence the name calculus of variations. For a given function Y the value of $J(y_\alpha)$ depends only on the variable α. We express this with

$$\Phi(\alpha) = J(y_\alpha) = \int_a^b f(x, y_\alpha(x), y_\alpha{}'(x))\, dx.$$

Now the function Φ can be represented by a Taylor expansion about the point $\alpha = 0$:

$$\Phi(\alpha) = \Phi(0) + \alpha\Phi'(0) + \text{ remainder.}$$

The linear part $\alpha\Phi'(0)$ is called the *first variation* of the variational integral.

Since y_0 is assumed to be the solution of the variational problem, we must have

$$\Phi(\alpha) = J(y_\alpha) \geq J(y_0) = \Phi(0).$$

That is, Φ has a relative minimum at the point $\alpha = 0$.

Now

$$\begin{aligned} \Phi'(0) = \int_a^b &[f_y(x, y_0(x), y_0'(x))Y(x) \\ &+ f_{y'}(x, y_0(x), y_0'(x))Y'(x)]\, dx. \end{aligned} \tag{2.3}$$

We add here parenthetically that in what follows we will use the following shorter expressions:

$$\begin{aligned} f_y(x, y_0(x), y_0'(x)) &= f_y^0(x) \quad (f_y \text{ along } y_0), \\ f_{y'}(x, y_0(x), y_0'(x)) &= f_{y'}^0(x) \quad (f_{y'} \text{ along } y_0). \end{aligned} \tag{2.4}$$

After integration by parts of the second term of the integrand in (2.3), we obtain

$$\int_a^b f_{y'}^0(x)Y'(x)\,dx = [f_{y'}^0(x)Y(x)]_a^b - \int_a^b \frac{d}{dx} f_{y'}^0(x)Y(x)\,dx. \tag{2.5}$$

Since $Y(a) = Y(b) = 0$, (2.3) becomes

$$\Phi'(0) = \int_a^b \left[f_y^0(x) - \frac{d}{dx} f_{y'}^0(x) \right] Y(x)\,dx = 0. \tag{2.6}$$

Now we employ a simple trick. Equation (2.6) is valid for each variation $\alpha Y(x)$ and therefore also valid for

$$\alpha Y(x) = \alpha(x-a)(b-x)\left[f_y^0(x) - \frac{d}{dx} f_{y'}^0(x) \right].$$

This special Y is smooth in $[a, b]$ since the solution y_0 is assumed to have continuous first, second and third derivatives. If we substitute this expression for Y in (2.6), we obtain

$$\int_a^b \left[f_y^0(x) - \frac{d}{dx} f_{y'}^0(x) \right]^2 (x-a)(b-x)\,dx = 0. \tag{2.7}$$

The integrand of this integral is non-negative on the interval $[a, b]$. Thus, (2.7) is possible only if

$$f_y^0(x) - \frac{d}{dx} f_{y'}^0(x) = 0 \quad \text{for all} \quad x \in (a, b). \tag{2.8}$$

This is the *Euler differential equation* for the solutions y_0 of the variational problem. Since y_0 does not appear in (2.8), we write out the more complete form

$$f_y(x, y_0(x), y_0'(x)) - \frac{d}{dx}[f_{y'}(\cdot, y_0, y_0')](x) = 0. \tag{2.9}$$

Some care is required here. Note that the expression $\frac{d}{dx} f_{y'}^0(x) = \frac{d}{dx}[f_{y'}(\cdot, y_0, y_0')](x)$ is not the partial derivative of $f_{y'}$ with respect to x, but rather the derivative with respect to x of the function $f_{y'}(x, y_0(x), y_0'(x))$ which depends only on the single variable x (cf. (2.4)). A smooth solution of the Euler equation is called an *extremal.* Since y_0 has continuous first, second and third derivatives, we can rewrite (2.9) to get

$$f_y - f_{y'x} - f_{y'y}y_0'(x) - f_{y'y'}y_0''(x) = 0. \tag{2.10}$$

In this equation the values x, $y_0(x)$ and $y_0'(x)$ appear as the arguments of the various partial derivatives. This expanded form of the Euler equation is not always the most favourable. Often there are only special solutions in closed form and it is advisable to use (2.8) rather than (2.10).

If the coefficient $f_{y'y'}$ of y_0'' in equation (2.10) is different from 0, then (2.10) can be solved for y_0'' so that it becomes a non-linear differential equation of second order.

Example

Find the shortest curve joining points $A = (a, y_a)$ and $B = (b, y_b)$ in the plane $(a < b)$ (cf. (1.1)).

Here the variational integral is

$$J(y) = \int_a^b \sqrt{1 + y'^2}\, dx.$$

Since the integrand here is not dependent on y, the solution y of the problem must satisfy the Euler equation

$$\frac{d}{dx}\frac{y'}{\sqrt{1 + y'^2}} = 0.$$

That is, $y'/\sqrt{1 + y'^2}$ must be a constant on $[a, b]$. But this means that y' is also a constant so that $y(x) = c_1x + c_2$.

The boundary conditions

$$y(a) = c_1a + c_2 = y_a, \quad y(b) = c_1b + c_2 = y_b,$$

are only fulfilled for

$$c_1 = \frac{y_b - y_a}{b - a}, \qquad c_2 = y_a - a\frac{y_b - y_a}{b - a}.$$

We conclude from this that, if there is a solution that possesses first, second and third derivatives, then that solution must be

$$y(x) = \frac{y_b - y_a}{b - a}(x - a) + y_a.$$

Clearly, y represents the curve joining A and B.

Up to the present we have assumed that y_0 be the absolute minimum of the extremal problem. But we can also derive the Euler equation for relative minima by the same method.

(2.11) Theorem

If a function y_0, having continuous first, second and third derivatives, is a strong or weak solution of the variational problem, then it must satisfy the Euler differential equation (2.8).

Proof

The function y_0 is a strong or weak solution of the variational problem if $J(y_0) < J(y)$ is valid for each admissible function y for which its d_0-distance or d_1-distance, respectively, from y_0 is sufficiently small. For admissible functions $y = y_\alpha$ of a one-parameter family $y_\alpha = y_0 + \alpha Y$, these distances may be evaluated in the following way:

$$\begin{aligned} d_0(y_\alpha, y_0) \leq d_1(y_\alpha, y_0) &= \sup_{x\in[a,b]} (|\alpha Y(x)| + |\alpha Y'(x)|) \leq |\alpha| M, \\ M &= \sup_{x\in[a,b]} (|Y(x)| + |Y'(x)|). \end{aligned}$$

For each variation αY there is, therefore, an $\alpha_0 > \alpha$ such that for $|\alpha| < \alpha_0$ the distances $d_0(y_\alpha, y_0)$ and $d_1(y_\alpha, y_0)$ are so small that $J(y_0) \leq J(y_\alpha)$ holds. The central assumption in the derivation of the Euler equation is therefore fulfilled so that the earlier derivation applies, word for word, in the present case.

2.2 INTEGRATION OF THE EULER EQUATION UNDER SPECIAL ASSUMPTIONS ON THE FUNCTION f

In general, it is not easy to integrate the Euler equation. Thus we want to develop explicit methods of solution for some classes of integrands f.

First case: The integrand f does not depend on y; i.e. $f_y = 0$. In this case, the Euler equation reduces to

$$\frac{d}{dx} f_{y'}(\cdot, y, y') = 0$$

from which it follows that $f_{y'}(x, y(x), y'(x)) = \text{constant} = c_1$. If this (implicit) differential equation can be solved for y', then we have

$$y'(x) = g(x; c_1)$$

and the extremals y of the problem can be found by integration:

$$y(x) = \int_a^x g(t; c_1)\, dt + c_2,$$

where c_1 and c_2 are integration constants.

Example

Consider the variational integral

$$J(y) = \int_a^b \sqrt{\frac{1 + y'^2(x)}{x}}\, dx \quad (0 < a < b).$$

Here the equation

$$f_{y'}(x, y(x), y'(x)) = \frac{y'(x)}{\sqrt{x(1 + y'^2(x))}} = c_1$$

can be solved for $y'(x)$:

if $c_1 = 0$, then $y' = 0$ so that $y(x) = c_2$;

if $c_1 \neq 0$, then $y'(x) = \pm\sqrt{\dfrac{c_1^2 x}{1 - c_1^2\, x}} = g(x, c_1)$.

The integration is simplified if we introduce the new independent variable u by means of the substitution

$$x = \hat{x}(u) = \frac{1}{c_1^2}\sin^2(u) = \frac{1}{2c_1^2}(1 - \cos^2(u)),$$

$$a = x(u_0), \quad 0 \leq u \leq \frac{\pi}{2}.$$

Then

$$\begin{aligned}\hat{y} = y(\hat{x}(u)) &= \pm\frac{2}{c_1^2}\int_{u_0}^{u}\sin^2(u)\,du + c_2^*\\ &= \pm\frac{1}{2c_1^2}(2u - \sin(2u)) + c_2, \quad 0 \leq u \leq \frac{\pi}{2}.\end{aligned}$$

For $c_1 \neq 0$ we obtain the extremal in the form of a parametric representation $(\hat{x}(u), \hat{y}(u))$ rather than in the form of $y = y(x)$. The curves with the stated parametric representation are *cycloids*.

Second case: The integrand f does not depend on x; i.e. $f_x = 0$. If y_0 is an extremal having continuous first, second and third derivatives, then the derivative of the following function H vanishes identically:

$$\begin{aligned}H(x) &= y_0'(x)f_{y'}^0(x) - f(x, y_0(x), y_0'(x)),\\ H'(x) &= y_0''f_{y'}^0(x) + y_0'(x)\frac{d}{dx}f_{y'}^0\\ &\quad - f_y^0(x)y_0'(x) - f_{y'}^0(x)y_0''(x) = 0.\end{aligned}$$

Thus $H(x) = c_1 =$ constant. Now consider the equation

$$y'f_{y'}(x, y, y') - f(x, y, y') = c, \tag{2.12}$$

which is an equation in y and y' associated with H. If we can solve this equation for y', then we will obtain a differential equation $y' = g(y, c)$ which can be solved by separation of variables.

Example

The problem (1.4) of Dido. Among the smooth functions y which are positive on $(0, \ell)$ and which vanish at the endpoints $x = 0$ and $x = \ell$, find that function which yields a minimum for

$$J(y) = -\int_0^\ell y(x)\sqrt{1 - y'^2(x)}\, dx.$$

The function H given above is, in this instance,

$$H(x) = \frac{y_0(x) y_0'^2(x)}{\sqrt{1 - y_0'^2(x)}} + y_0(x)\sqrt{1 - y_0'^2(x)} = c_1.$$

If we solve for $y_0'(x)$ we find that

$$\begin{aligned} c_1 y_0'(x) &= \pm\sqrt{c_1^2 - y_0^2(x)} \quad \text{if} \quad c_1 \neq 0, \\ y_0(x) &= 0 \quad \text{if} \quad c_1 = 0. \end{aligned}$$

If $c_1 \neq 0$, then separating variables and making the substitution $\hat{y}(u) = c_1 \sin(u)$ leads to the expression

$$\begin{aligned} \pm \int dx = \pm x + \hat{c}_2 = \int \frac{c_1\, dy}{\sqrt{c_1^2 - y^2}} &= \int \frac{c_1^2 \cos(u)}{c_1 \cos(u)}\, du \\ &= c_1 u + c_2^*. \end{aligned}$$

Therefore, $x = c_1 u + \tilde{c}_2$ and $u = (1/c_1)x + c_2$, respectively, with integration constants c_1 and c_2. The functions

$$\begin{aligned} y(x) &= 0 \text{ and} \\ y(x) &= c_1 \sin\left(\frac{x}{c_1} + c_2\right) \end{aligned}$$

are, therefore, the extremals of the variational problem. The function $y(x) = 0$, however, does not satisfy the conditions set down in the variational problem. The boundary conditions $y(0) = y(\ell) = 0$ are satisfied by

$$y(x) = \frac{1}{\pi}\ell \sin\left(\frac{\pi}{\ell}x\right).$$

The values of this function are positive for $x \in (0, \ell)$; thus y is an admissible function.

Finally, we must consider what this extremal means for the problem of Dido. The maximum of $I(G^+)$ (cf. (1.9)) is clearly the minimum of $J(y)$ with ℓ replacing $L/2$.

In the integral $I(G^+)$ y appears as a function of arc length rather than of the abcissa x. An extremal of $I(G^+)$ is then given by

$$y(s) = \frac{\ell}{\pi} \sin\left(\frac{\pi}{\ell} s\right).$$

The parameter s is precisely the length of arc. Suppose the condition $x'^2(s) + y'^2(s) = 1$ is satisfied. Then, from this condition, we can determine $x(s)$

$$\begin{aligned} x(s) &= \int_0^s \sqrt{1 - y'^2(\sigma)}\, d\sigma \\ &= \int_0^s \sin\left(\frac{\pi}{\ell}\sigma\right) d\sigma = \frac{\ell}{\pi} \cos\left(\frac{\ell}{\pi} s\right) - \frac{\ell}{\pi}. \end{aligned}$$

For $s \in [0, \ell]$ the parametric representation $(x(s), y(s))$ determines a half-circle with radius $r = \frac{\ell}{\pi}$ and centre at $M = (\frac{\ell}{\pi}, 0)$.

By this means it is shown that the half-circle with the radius $r = \frac{L}{2\pi}$ is the solution of the problem of Dido.

Third case: The integrand f is a linear function of y' so that $f_{y'y'} = 0$. In this case, f has the form

$$f(x, y, y') = A(x, y) = B(x, y)y'$$

and has continuous first, second and third partial derivatives. Here the Euler equation is

$$\begin{aligned} &\frac{d}{dx} B(\cdot, y)(x) - Ay(x, y(x)) - By(x, y(x))y'(x) \\ &\qquad = B_x(x, y(x)) - A_y(x, y(x)) = 0. \end{aligned}$$

This is not a differential equation but rather an equation to be satisfied by y. (In spite of this, we will still refer to this equation as the Euler equation, since such is the usual convention.) In this case we cannot expect that a solution will exist for arbitrary boundary conditions y_a and y_b. An important special case arises when $B_x - B_y$ is everywhere zero. Then the variation integral is independent of the path of integration, and every admissible function is a solution of the variational problem (cf. Section 6.4).

2.3 FURTHER EXAMPLES AND PROBLEMS

We show next a simple example from economics: *the Ramsey growth model.* For the production of a homogeneous product, an amount of capital K is invested. In order to make the model simple, we assume that the production Y does not depend on other values, like the available work force, for example. The production Y is, therefore, only a function of the available capital

$$Y = \psi(K).$$

Of the production Y a part C will be used up (consumed) and the remainder, $I = Y - C$, will be invested so that a change in capital results. Thus,

$$I = \frac{dK}{dt}, \quad t = \text{ time.}$$

We will require that C have only positive values. I, on the other hand, may have positive and negative values. $I < 0$ implies that the capital is being reduced because of consumption. These relationships are expressed in the equations

$$Y = \frac{dK}{dt} + C = \psi(K), \quad C = \psi(K) - \frac{dK}{dt}.$$

The question now arises, How shall the distribution of consumption and investment at any time t be controlled so that a certain value, which we will call the (total) product ν, is maximized? The total product ν is given by the integral of the instantaneous product $N(t)$ which depends on the consumption $C(t)$ at the time t

$$N(t) = N(C(t)),$$

$$\nu = \int_{t_0}^{t_1} N(C(t))\,dt = \int_{t_0}^{t_1} N\left(\psi(K(t)) - \frac{dK(t)}{dt}\right) dt \to \max.$$

The times t_0 and t_1 are considered to be fixed points in time; t_1 is the so-called planning horizon and may have any value $> t_0$ including the limiting 'value' $t = \infty$. As soon as the functions Ψ and N are determined, we can begin the calculations. A simplified (but sensible) possibility is

$$\Psi(K) = bK \quad \text{(production } Y \text{ proportional to capital)},$$
$$N(C) = -a(C - C^*)^2,$$

where C^*, a and b are constants with $a, b > 0$.

We can now write the variation integral

$$-\nu(K) = \int_{t_0}^{t_1} a\left(bK(t) - \frac{dK(t)}{dt} - C^*\right) dt \to \min. \qquad (2.13)$$

The integrand does not depend explicitly on the independent variable t. We have here an example of the second case (cf. Section 2.2), so that an extremal $K(t)$ of the problem must satisfy the condition

$$\frac{dK}{dt} 2a\left(bK - \frac{dK}{dt} - C^*\right) + a\left(bK - \frac{dK}{dt} - C^*\right)^2 = c_1 = \text{constant}.$$

If we solve this equation for $\frac{dK}{dt}$ we obtain

$$\frac{dK}{dt} = \pm\sqrt{(bK - C^*)^2 - c_2}, \qquad c_2 = \frac{c_1}{a}.$$

Separation of variables and (if $c_2 \neq 0$) the substitution

$$u = \frac{bK - C^*}{\sqrt{| c_2 |}}$$

leads to

$$\pm t + c_3 = \int \frac{dK}{bK + C^*} = \frac{1}{b} \ln | bK(t) - C^* | \quad \text{if} \quad c_2 = 0,$$

$$\pm t + c_3 = \int \frac{dK}{\sqrt{(bK - C^*)^2 - c_2}}$$

$$= \begin{cases} \frac{1}{b}\int \frac{du}{\sqrt{u^2-1}} & = \frac{1}{b}\cosh^{-1}(u) + c_4 \quad \text{if} \quad c_2 > 0, \\ \frac{1}{b}\int \frac{du}{\sqrt{u^2+1}} & = \frac{1}{b}\sinh^{-1}(u) + c_4 \quad \text{if} \quad c_2 < 0. \end{cases}$$

If we now replace u with its representation in terms of K, we obtain finally the extremals

$$K(t) = K^* + c_5 e^{\pm bt} \qquad \text{if} \quad c_2 = 0,$$

$$K(t) = K^* \pm \frac{\sqrt{c_2}}{b}\sinh(bt + c_5) \quad \text{if} \quad c_2 < 0,$$

$$K(t) = K^* \pm \frac{\sqrt{c_2}}{b}\cosh(bt + c_5) \quad \text{if} \quad c_2 > 0,$$

where $K^* = \frac{C^*}{b}$ and c_2 and c_5 are constants of integration. We will not include an interpretation of these results here.

The next example is a special case of problem (1.1). We seek the *shortest curve* between two points which lie on a sphere. The sphere $K(O, r)$ with centre at the origin and having radius r may be represented locally by means of parameters which correspond to longitude x and latitude y on the sphere:

$$r(\cos x \cos y, \sin x \sin y, \sin y),$$

$$x \in (0, 2\pi), \qquad y \in (-\frac{\pi}{2}, \frac{\pi}{2}).$$

We consider curves on the sphere which cut each meridian at only one point. For these curves we can choose the longitude x as a parameter. Then the latitude y of a point on the curve will be described by a function y of the independent variable x. For $x \in [a, b]$,

$$x \in [a, b] \mapsto r(\cos x \cos y(x), \sin x \cos y(x), \sin y(x))$$

gives a parametric representation of the curve. If y is smooth, then the length of the curve is obtained from

$$L(y) = r\int_{x_0}^{x_1} \sqrt{\cos^2 y(x) + y'^2(x)}\, dx. \tag{2.14}$$

The integrand of this variational integral is independent of x. Consequently, the function H is constant along an extremal y (cf. Section 2.2, second case) and

$$-\cos^2 y(x) = c\sqrt{\cos^2 y(x) + y'^2(x)}, \quad c = \text{ const.} \tag{2.15}$$

For $c = -1$ the equator ($y = 0$) is the solution. From (2.15) it follows that

$$c^2 y'^2(x) = \cos^4 y(x) - c^2 \cos^2 y(x).$$

Separation of variables and the substitution $t = \tan y$ leads to the equation

$$\begin{aligned}\pm(x + x_0) &= \int \frac{c\,dy}{\cos y\sqrt{\cos^2 y - c^2}} = \int \frac{c\,dt}{\sqrt{(1-c^2) - c^2t^2}} \\ &= \arcsin \frac{ct}{\sqrt{1-c^2}} \qquad (0 < c^2 < 1).\end{aligned}$$

Thus,

$$\tan y(x) = \frac{\sqrt{1-c^2}}{c}\sin(\pm x + x_0)$$

and

$$y(x) = \arctan\left[\frac{\sqrt{1-c^2}}{c}\sin(\pm x + x_0)\right].$$

These functions y represent great circles on the sphere.

Finally, we want to consider the problem (1.2) of *the smallest surface of revolution*:

$$\begin{aligned}J(y) = \int_a^b y(x)\sqrt{1 + y'^2(x)}\,dx &\to \min, \quad y(x) > 0, \\ x \in [a, b], \quad y(a) = y_a \quad &\text{and} \quad y(b) = y_b.\end{aligned}$$

We will determine the extremal according to the formula of Section 2.2. Then we will show that for the prescribed boundary values there can be one, two or no extremals.

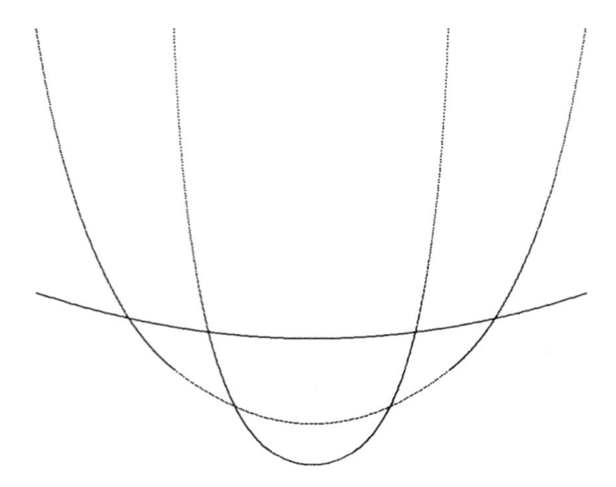

Fig. 2.1. Catenaries.

The integrand here is independent of x. Thus, from the Euler equation we must have

$$y(x)\sqrt{1+y'^2(x)} - y(x)\frac{y'^2(x)}{\sqrt{1+y'^2(x)}} = c = \frac{y(x)}{\sqrt{1+y'^2(x)}} > 0.$$

Solve for y' to obtain

$$y'(x) = \sqrt{\frac{1}{c^2}y^2(x) - 1}.$$

Then separating variables leads to the extremals

$$y(x) = \frac{1}{c}\cosh(cx+d).$$

Here $c > 0$ and d are constants of integration. These curves are called *catenaries* (cf. Fig. 2.1). At the left endpoint set $A = (a, y_a) = (0, 1)$. Also let $b = 1$. We will not assign a particular value to y_b but rather investigate which values $y_b > 0$ may have here. For the extremals through $A = (0, 1)$ we have

$$1 = \frac{1}{c}\cosh d, \quad y(x) = \frac{\cosh(d + x\cosh d)}{\cosh d},$$

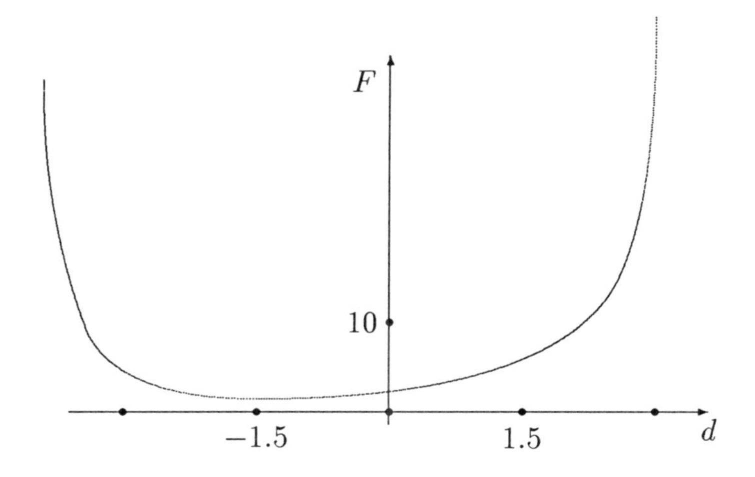

Fig. 2.2. Graphical determination of minimum for F.

where d may be determined from the boundary condition

$$y_b = \frac{\cosh(d + \cosh d)}{\cosh d} = F(d). \tag{2.16}$$

Since (2.16) is generally not solvable for d in a simple way, we provide a graphic interpretation. Since $F(d) \to \infty$ as $d \to \pm\infty$, it is clear that F must have a minimum. The minimum value for F is a positive number $y_{\min} = \min F(d)$. If $y_b < y_{\min}$, then there is no solution d for the equation (2.16) and, therefore, no extremal through $A = (0, 1)$ and $B = (1, y_b)$. If, on the other hand, $y_b = y_{\min}$, then there is an extremal through A and B. If $y_b > y_{\min}$, then at least two extremals through A and B exist.

The above remarks summarize the qualitative investigation. If one wants to find the extremals through A and B for the special problem where $y_b > y_{\min}$, then one can find an approximation for the parameter d satisfying (2.16) from the diagram shown in Fig. 2.2.

Exercises

(2.17) In the study of the propagation of light in a material where the speed ψ of the light depends on position, the resulting variational problem has the integrand

$$f(x, y, y') = \frac{\sqrt{1 + y'^2}}{\psi(x, y)}.$$

Find the Euler equation under the assumption that ψ is partially differentiable.

(2.18) Find the extremals of the variational problem having the integrand

$$f(x, y, y') = x^4 y'^2.$$

Is there an extremal satisfying the boundary conditions

$$(a, y_a) = (-1, -2), \qquad (b, y_b) = (1, 1)?$$

(2.19) Find the extremals of the Ramsey model if N and ψ are given, respectively, by

(i) $N(C) = \alpha(C - C_0), \psi(K) = C_0 - \beta(K - K^*), \alpha, \beta, C_0 > 0,$
(ii) $N(C) = -\alpha(C - C_0)^2, \psi(K) = \beta\sqrt{K}, \alpha, \beta, C_0 > 0.$

2.4 NEW ADMISSIBILITY CONDITIONS AND THE LEMMA OF DU BOIS–REYMOND

There are variational problems in which one must deal with 'solution curves' which have corners. These are curves for which the derivative has 'jumps'. Recall the example of sailing against the wind on a river where the flow is zero. For this example, the optimizing path while tacking has corners.

Up to this point we have only permitted differentiable functions to be admissible. So it can happen under these restrictions that a variational problem has no solution since too few functions are admissible. We will, therefore, weaken the admissibility condition. To this end, we need the concept of a piecewise continuous and piecewise smooth function.

(2.20) Definition
We say that a function $F : [a, b] \mapsto \mathbb{R}$ is *piecewise continuous* if there are finitely many points $x_1, \cdots, x_r$ with $a < x_1 < x_2 < \ldots < x_r < b$ such that F is continuous in the open intervals (a, x_1), (x_1, x_2), $\ldots$, (x_r, x_b) and, furthermore, that each of the following limits exists:

$$\lim_{x \to a^+} F(x), \quad \lim_{x \to x_i^-} F(x), \quad \lim_{x \to x_i^+} F(x), \quad \lim_{x \to b^-} F(x), \quad i = 1, \ldots, r.$$

(2.21) Definition
A function $y : [a, b] \mapsto \mathbb{R}$ is called *piecewise smooth* if y is continuous and if there are finitely many points $x_1, x_2, \ldots, x_n$ with $a < x_1 < x_2 < \cdots < x_n < b$ and such that y is smooth on each of the intervals $[a, x_1]$, $[x_1, x_2]$, $\ldots$, $[x_{n-1}, x_n]$, $[x_n, b]$. Here we define the derivative at the left and right endpoints, respectively, of the intervals of definition $[x_k, x_{k+1}]$ to be

$$y'(x_k) = y'_r(x_k) = \lim_{x \to x_k^+} \frac{y(x) - y(x_k)}{x - x_k} \text{ (right-hand derivative),}$$

$$\begin{aligned} y'(x_{k+1}) &= y'_l(x_{k+1}) \\ &= \lim_{x \to x_{k+1}^-} \frac{y(x) - y(x_{k+1})}{x - x_{k+1}} \text{ (left-hand derivative).} \end{aligned}$$

We call x_k a *corner* of y if $y'_l(x_k) \neq y'_r(x_k)$.

In what follows we will investigate a variational problem with a variational integral J as before, but with the following altered conditions for the set of admissible functions.

(2.22) Weakened conditions for admissible functions
The set of admissible functions consists of all piecewise smooth functions $y : [a, b] \mapsto \mathbb{R}$ which have at the endpoints a and b the prescribed values y_a and y_b; $y(a) = y_a$, $y(b) = y_b$.

Before continuing to the next section, in which there will appear a variational problem whose solution has corners and which requires that necessary conditions be satisfied at the endpoints, we want to prove a needed lemma.

(2.23) Lemma of du Bois–Reymond

Let F be piecewise continuous on $[a,b]$ and such that for all piecewise smooth functions $Y : [a,b] \mapsto \mathbb{R}$ with the boundary conditions $Y(a) = Y(b) = 0$ we have

$$\int_a^b F(x)Y'(x)\,dx = 0.$$

Then F is constant on $[a,b]$.

Proof

We introduce the mean value c of F

$$c = \frac{1}{b-a}\int_a^b F(x)\,dx$$

and observe that the function Y_0 given by

$$Y_0(x) = \int_a^x [F(u) - c]\,du$$

is defined, is piecewise smooth and satisfies the boundary conditions

$$Y_0(a) = 0 = Y_0(b).$$

For all $x \in [a,b]$ where F is continuous, i.e. for all x except for finitely many values x_i, it holds that $Y_0'(x) = F(x) - c$. Thus, according to the assumption of the lemma,

$$\begin{aligned} 0 &= \int_a^b Y_0'(x)F(x)\,dx = \int_a^b (F(x)-c)F(x)\,dx \\ &= \int_a^b [(F(x)-c)F(x) - c(F(x)-c)]\,dx \\ &= \int_a^b [F(x)-c]^2\,dx = 0. \end{aligned}$$

(From the boundary conditions, $cY_0(b) = \int_a^b c(F(x)-c)\,dx = 0$.)

From the above discussion we conclude that if F is continuous at the point x, then $F(x) = c$ must hold. F was assumed to be piecewise continuous. But this means that the limits

$$\lim_{x \to x_i^-} F(x) \quad \text{and} \quad \lim_{x \to x_i^+} F(x)$$

exist and are equal to c. Thus at the point x_i, F can be defined uniquely by $F(x_i) = c$.

We conclude that for all $x \in [a, b]$, $F(x) = c =$ constant is valid.

2.5 THE ERDMANN CORNER CONDITIONS

We begin with the approach which in Section 2.1 led us to the Euler equation. Now, however, the admissible functions need only be piecewise smooth so that the necessary condition derived from the first variation has a different form.

In this section, y_0 is a piecewise smooth function and a strong or weak solution of the variational problem with the weakened admissibility conditions (2.22). We consider a family $y_\alpha = y_0 + \alpha Y$ of piecewise smooth admissible functions. Here α is the parameter of the family, $|\alpha| < \alpha_0$, and Y is an arbitrary, piecewise smooth function satisfying the boundary conditions $Y(a) = Y(b) = 0$. The corners of y_α are at the points $x_1, x_2, \ldots, x_m$. As with the derivation of y_α, we set $\Phi(\alpha) = J(y_\alpha)$ and derive from the requirement $\Phi'(0) = 0$ the expression

$$\Phi'(0) = \int_a^b [f_y^0(x)Y(x) + f_{y'}^0(x)Y'(x)]\,dx = 0. \tag{2.24}$$

We recall in this connection that

$$f_y^0(x) = f_y(x, y_0(x), y_0'(x))$$

and

$$f_{y'}^0(x) = f_{y'}(x, y_0(x), y_0'(x)).$$

Integration of (2.5) by parts is not permissible here for the reason that $f^0_{y'}(x) = f_{y'}(x, y_0(x), y_0'(x))$ is now only piecewise continuous. But we can integrate the first term of (2.24) by parts to obtain

$$\int_a^b f^0_y(x)Y(x)\,dx = \sum_{i=1}^{m+1}\left[\int_{x_{i-1}}^x f^0_y(u)\,du\,Y(x)\right]_{x=x_{i-1}}^{x_i}$$
$$-\int_a^b\left[Y'(x)\int_a^x f^0_y(u)\,du\right]dx \quad (x_0 = a,\ x_{m+1} = b). \tag{2.25}$$

Since $Y(a) = Y(b) = 0$, the first and last terms of the sum are both equal to zero. The remaining terms of the sum vanish in pairwise fashion. If we substitute the result to (2.25) in (2.24) we obtain

$$0 = \int_a^b\left[f^0_{y'}(x) - \int_a^x f^0_y(u)\,du\right]Y'(x)\,dx, \tag{2.26}$$

which is valid for all piecewise smooth functions Y with $Y(a) = Y(b) = 0$.

The function $F(x) = f^0_{y'}(x) - \int_a^x f^0_y(u)\,du$ is piecewise continuous on $[a, b]$. Thus the assumptions of the Lemma of du Bois–Reymond (2.23) are satisfied and it follows that

$$F(x) = f^0_{y'}(x) - \int_a^x f^0_y(u)\,du = \text{ constant}. \tag{2.27}$$

If the left-hand derivative $y'_{0l}(x_i)$ of y_0 at the point x_i does not agree with the right-hand derivative $y'_{0r}(x_i)$ of y_0, then we form the difference

$$F(x_2) - F(x_1) = f^0_{y'}(x_2) - f^0_{y'}(x_1) - \int_{x_1}^{x_2} f^0_y(u)\,du = 0,$$

where $x_1 < x_i < x_2$. If we let x_1 and x_2 tend to x_i, the integral will vanish since the integrand is a bounded function. This leads directly to the condition

$$\begin{aligned}\lim_{x\to x_i^+} f^0_{y'}(x) &= f_{y'}(x_i, y_0(x_i), y'_{0r}(x_i))\\ = \lim_{x\to x_i^-} f^0_{y'}(x) &= f_{y'}(x_i, y_0(x_i), y'_{0l}(x_i)),\end{aligned} \tag{2.28}$$

which forms the first *Erdmann corner condition.* This corner condition can be expressed briefly: $f^0_{y'}$ is a continuous function on $[a, b]$.

After stating a first condition, we naturally expect a second. We provide it here without derivation

$$f^0 - f^0_{y'} y_0' \quad \text{is continuous on } [a, b], \tag{2.29}$$

where $f^0(x) = f(x, y_0(x), y_0'(x))$. Still more explicitly, the corner condition requires that, for all endpoints x_i of y_0,

$$\begin{aligned} &f(x_i, y_0(x_i), y_{0l}'(x_i)) - f_{y'}(x_i, y_0(x_i), y_{0l}'(x_i)) y_{0l}'(x_i) \\ &= f(x_i, y_0(x_i), y_{0r}'(x_i)) - f_{y'}(x_i, y_0(x_i), y_{0r}'(x_i)) y_{0r}'(x_i). \end{aligned}$$

Proof of the above will be provided in Section 9.3.

Example

Consider the variational integral

$$J(y) = \int_a^b (y'^2(x) - 1)^2 \, dx.$$

Before we routinely attempt a general solution of this problem it would be profitable to consider the problem heuristically. Clearly, the integral is zero when $y' = \pm 1$ and greater than zero otherwise. Consequently, those admissible functions whose derivatives (piecewise) are $+1$ or -1 are solutions of the problem.

We will forget these observations now, however, in order to study the meaning of the necessary conditions of the first variation for this example.

The Euler equation for this problem is

$$\frac{d}{dx} f^0_{y'}(x) = 0 = \frac{d}{dx}[4y'(y'^2 - 1)].$$

Thus the condition that $4y'(x)(y'^2(x) - 1)$ be constant must be satisfied by the extremals. We can conclude that $y'(x)$ is constant and that $y = c_1 x + c_2$. We need to check now whether corners can occur. To this end (cf. (2.28) and (2.29)), we investigate the expressions

$$f_{y'}(x, y, y') = 4y'(y'^2 - 1)$$

and

$$\begin{aligned} f(x, y, y') - f_{y'}(x, y, y')y' &= (y'^2 - 1)^2 - 4y'^2(y'^2 - 1) \\ &= (y'^2 - 1)(-3y'^2 - 1). \end{aligned}$$

The corner conditions require that for arbitrary x and y we must have

$$f_{y'}(x, y, -1) = f_{y'}(x, y, 1)$$

and

$$f(x, y, -1) - f_{y'}(x, y, -1)(-1) = f(x, y, 1) - f_{y'}(x, y, 1)1.$$

Clearly we must reckon here with corners at which the left- and right-handed derivatives can be ± 1. For the prescribed boundary values y_a and y_b,

$$y(x) = \frac{y_b - y_a}{b - a}(x - a) + y_a$$

is an extremal and an admissible function, but in many cases it is not a solution of the variational problem. For many boundary values y_a, y_b (which ones? cf. Fig. 2.3) there are admissible functions y which are defined piecewise by

$$y(x) = \pm x + c, \qquad c = \text{ constant},$$

which satisfy the corner conditions at their corners and which also satisfy the Euler equation elsewhere. These 'zig-zag' functions are solutions.

Note: From the integral relation (2.27) it follows that if y_0 is smooth for $x_0 \in (a, b)$, then $\int_a^x f_y^0(u)\, du$ is differentiable with

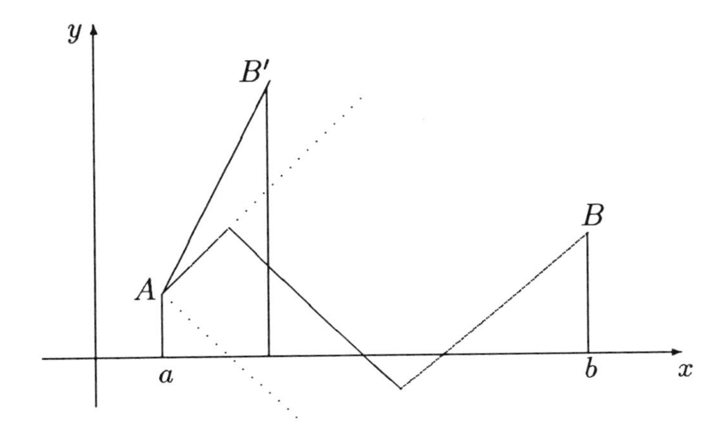

Fig. 2.3. Possible solutions with corners.

respect to x at x_0. Therefore, $f^0_{y'}$ is also differentiable and we may write (cf. (2.8))

$$f^0_y(x_0) - \frac{d}{dx} f^0_{y'}(x_0) = 0.$$

This proves a sharper result than we obtained in Theorem (2.11). (In that instance, it was necessary to assume that y_0 has continuous first, second and third derivatives.)

(2.30) Theorem
A piecewise smooth solution y_0 of the variational problem satisfies the Euler equation (2.8) at those points where y_0 has a continuous derivative.

If we write $\frac{d}{dx} f^0_{y'}(x)$ as the limit of a difference quotient and make use of the differentiability of f, then we can prove the following theorem.

(2.31) Theorem

If a smooth function $y_0 : [a, b] \mapsto \mathbb{R}$ *satisfies the Euler equation* (2.8) *and if, for* $x \in (a_1, b_1) \in [a, b]$, $f_{y'y'}(x) \neq 0$, *then in the interval* (a_1, b_1) y_0 *has continuous first and second derivatives.*

Exercise

Consider the integral

$$\int_{x_1}^{x_2} \left[\frac{y(x)}{y'^2(x)} + y'^2(x) \right] dx, \quad y'(x) \neq 0.$$

Show that for this problem corners are possible. (*Hint*: The corner conditions are satisfied if the substitution $y'_r(x_0) = -y'_l(x_0)$ is made.)

3

Sufficient conditions for variational problems with convex integrands

---

We are now in the position to name the candidates for the solution from among the admissible functions, i.e. the extremals which for all $x \in [a, b]$ satisfy either the corner condition or the Euler equation.

If one knows that the variational problem has a solution and if there is only one candidate for the solution, then this y_0 is, of course, the solution to the problem. This situation does arise occasionally among variational problems deriving from certain applications and for which the person who formulated the problem has further information.

If such good fortune is not ours, however, we still want to be able to decide whether a particular candidate is a solution or not. We recall that for functions of one variable there is no simple general criterion for extrema but only either necessary or sufficient conditions. So is it also for variational problems. In the following paragraphs we will derive sufficient conditions which are easy to understand and are important for applications. First, however, we must make some assumptions – assumptions which will fail to be satisfied by many examples.

3.1 CONVEX FUNCTIONS

Convex functions play an important role in connection with different extreme value and optimization problems. In this section, we will derive those properties of convex integrands which are of interest to the calculus of variations. Finally – with attention to practical applications – we will provide a simple test with which one can tell whether a given function is convex or not. The test will be demonstrated with an example.

(3.1) Definition
A set M of numbers or of points of $\mathbb{R}^n$ ($n = 1, 2, 3, \ldots$) is said to be *convex* if, for any two elements $u_1, u_2 \in M$, all numbers or points which lie on the segment joining u_1 and u_2 belong to M. That is, M is convex if, for all $u_1, u_2 \in M$ and all $t \in [0, 1]$,

$$u_1 + t(u_2 - u_1) = tu_2 + (1 - t)u_1 \in M.$$

The space $\mathbb{R}^n$, quadrilaterals, the plane, and the sphere in 3-space are all examples of convex sets. The interior of the curve in Fig. 1.4 is a set which is not convex.

(3.2) Definition
A function f which is defined on a convex set is said to be *convex* if for all $u_1, u_2 \in M$ and all $t \in (0, 1)$ it holds that

$$f(tu_2 + (1 - t)u_1) \leq tf(u_2) + (1 - t)f(u_1). \qquad (3.3)$$

One can see from its graph whether a function is convex or not. An easy calculation shows that f is convex if the set of points lying 'above' the graph of f, i.e. the set $\{(u, z) \in M \times \mathbb{R} : z \geq f(u)\}$, is also convex.

Our sufficient conditions have to do with a generalization of the following theorem.

(3.4) Theorem
Let f be a convex, smooth function of a variable x with an interval of definition M. If $f'(x_0) = 0$ for $x_0 \in M$, then x_0 provides a minimum for f.

Proof
For an arbitrary $x \in M$, Taylor's formula gives

$$\begin{aligned} f(tx + (1-t)x_0) &= f(x_0 + t(x - x_0)) \\ &= f(x_0) + f'(x^*)t(x - x_0), \end{aligned}$$

where x^* is a value between x_0 and x. From the convexity of f it follows that

$$f(tx + (1-t)x_0) \leq tf(x) + (1-t)f(x_0).$$

A comparison of the above now yields

$$f(x_0) + f'(x^*)t(x - x_0) \leq tf(x) + f(x_0) - tf(x_0), \quad t \in (0,1)$$

and

$$f(x_0) \leq f(x) - f'(x^*)(x - x_0).$$

Since f' is continuous, we may take the limit as $t \to 0$, i.e. as $x^* \to x_0$; thus

$$f(x_0) \leq f(x) - f'(x_0)(x - x_0). \tag{3.5}$$

If now $f'(x_0) = 0$, it follows that $f(x_0) \leq f(x)$, which was to be shown.

But how can one determine whether a function f is convex or not, if one does not know the graph of the function exactly – or especially if f depends on two or more variables? The following theorem provides a criterion.

(3.6) Theorem
Let the function $f : M \subset \mathbb{R}^2 \mapsto \mathbb{R}$ have continuous first and second partial derivatives. Then f is convex if and only if for all $(x,y) \in M$

$$\begin{aligned} &f_{xx}(x,y)\, f_{yy}(x,y) - f_{xy}^2(x,y) \geq 0, \\ &f_{xx}(x,y) \geq 0, \quad \textit{and} \quad f_{yy}(x,y) \geq 0. \end{aligned} \tag{3.7}$$

The conditions given in (3.7), as is well known, are the conditions that the Hesse matrix for f be positive semi-definite; i.e. that the quadratic expression

$$Q(x, y; v, w) = f_{xx}(x, y)v^2 + 2f_{xy}(x, y)vw + f_{yy}(x, y)w^2 \quad (3.8)$$

be equal to or larger than zero for all v, $w \in \mathbb{R}$.

Proof

1. We show first that if f is convex, then for all $v, w \in \mathbb{R}$ and all $(x, y) \in M$, it follows that $Q(x, y; v, w) \geq 0$. Thus, let $u_1 = (x_1, y_1)$ and $u_2 = (x_2, y_2)$ be two arbitrary points in M and let $t \in (0, 1)$. Then, since f is convex,

$$f(tu_2 + (1 - t)u_1) \leq tf(u_2) + (1 - t)f(u_1)$$

and, according to the Taylor formula,

$$\begin{aligned} f(tu_2 + (1 - t)u_1) &= f(u_1 + t(u_2 - u_1)) \\ &= f(u_1) + \text{ grad } f(u^*) \cdot t(u_2 - u). \end{aligned}$$

Thus it follows that $f(u_2) \geq f(u_1) +$ grad $f(u^*) \cdot (u_2 - u_1)$. Here u^* is a point on the segment joining u_1 and $u_1 + t(u_2 - u_1)$. If we let $t \to 0$ we obtain the inequality

$$f(u_2) \geq f(u_1) + \text{ grad } f(u_1) \cdot (u_2 - u_1), \quad (3.9)$$

which is the analogue of (3.5).

In a similar way the Taylor expansion for $f(u_2)$ gives

$$f(u_2) = f(u_1) + \text{grad } f(u_1) \cdot (u_2 - u_1) + \tfrac{1}{2} Q(x^*, y^*; x_2 - x_1, y_2 - y_1),$$

where (x^*, y^*) is a point on the segment joining u_1 and u_2. Using (3.9), we can easily see that

$$Q(x^*, y^*; x_2 - x_1, y_2 - y_1) \geq 0. \quad (3.10)$$

If we now hold $u_1 = (x, y)$ fixed and set $u_2 = u_1 + \alpha z$ with $\alpha > 0$ and $z = (v, w) \in \mathbb{R}^2$, then it follows from (3.10) that

$$Q(x^*, y^*; v, w) \geq 0$$

(cf. (3.8) for the definition of Q). If we let $\alpha \to 0$ so that $(x^*, y^*) \to (x, y)$, we obtain

$$Q(x, y; v, w) \geq 0 \quad \text{for all} \quad (x, y) \in M, \quad (v, w) \in \mathbb{R}^2.$$

We conclude, therefore, that each convex function f of two variables satisfies the conditions of (3.7).

2. We want to show next that the convexity of f follows from the condition (3.7). To this end, let $u_1 = (x_1, y_1)$ and $u_2 = (x_2, y_2)$ be arbitrary points of M and let $t \in (0, 1)$. From the Taylor formula for $f(u_1)$ and $f(u_2)$ about the point

$$u = tu_2 + (1 - t)u_1 = u_1 + t(u_2 - u_1) = u_2 + (1 - t)(u_1 - u_2),$$

we have

$$\begin{aligned} f(u_1) = f(u) &- t \operatorname{grad} f(u) \cdot (u_2 - u_1) \\ &+ \tfrac{1}{2} t^2 Q(x^*, y^*; x_2 - x_1, y_2 - y_1) \end{aligned} \tag{3.11}$$

and

$$\begin{aligned} f(u_2) = f(u) &+ (1 - t) \operatorname{grad} f(u) \cdot (u_2 - u_1) \\ &+ \tfrac{1}{2}(1 - t)^2 Q(x^{**}, y^{**}; x_2 - x_1, y_2 - y_1), \end{aligned} \tag{3.12}$$

where (x^*, y^*) and (x^{**}, y^{**}) are points on the line segments joining u to u_1 and u_2, respectively.

If we multiply (3.11) with $(1 - t)$ and (3.12) with t, and then add, recalling that $Q \geq 0$, we obtain

$$(1 - t)f(u_1) + t(f(u_2), \geq f(u) = f(tu_2 + (1 - t)u_1).$$

But this implies that f is convex, which was to be shown.

Two examples

1. The integrand for the variational integral of the Ramsey model is

$$f(K, K') = a(bK - K' - C^*)^2, \quad a, \; b > 0.$$

Using the criteria of Theorem (3.6) we can show that f is a convex function on $M = \mathbb{R}^2$:

$$f_{KK} = 2ab^2 > 0, \quad f_{K'K'} = 2a > 0, \quad f_{KK'} = -2ab,$$

$$f_{KK}f_{K'K'} - f_{KK'}^2 = 0.$$

2. The $f(y, y') = y\sqrt{1 + y'^2(x)}$, which appears as the integrand for the problem of minimal surfaces of rotation (cf. (1.2)) is not convex for $M = \mathbb{R}^2$. Here

$$f_{yy}(y, y') = 0, \quad f_{yy'}(y, y') = \frac{y'}{\sqrt{1 + y'^2(x)}} \neq 0 \quad \text{if} \quad y' \neq 0.$$

Thus,

$$f_{yy}f_{y'y'} - f_{yy'}^2 = -\frac{y'^2}{1 + y'^2(x)} < 0 \quad \text{if} \quad y' \neq 0.$$

3.2 A SUFFICIENT CONDITION

We may now proceed to find an estimate for the difference $J(y) - J(y_0)$; from this we may obtain a sufficient condition. Of basic importance here is the inequality (3.9). In addition to the usual assumptions about the integrand f of the variational problem, we also assume that for each x

$$\begin{aligned} &\text{the set } M_x = \{(y, y') : (x, y, y') \text{ is in} \\ &\qquad \text{the domain of definition of } f\} \text{ is convex} \end{aligned} \tag{3.13}$$

and

$$f(x, y, y'), \text{ as a function on } M_x, \text{ is convex.} \tag{3.14}$$

Then for all points (y_0, y_0') and (y, y') from M_x it is valid that

$$\begin{aligned} f(x, y, y') \geq (x, y_0, y_0') &+ f_y(x, y_0, y_0')(y - y_0) \\ &+ f_y'(x, y_0, y_0')(y' - y_0'). \end{aligned} \tag{3.15}$$

Now let y and y_0 be two piecewise smooth admissible functions, and let the corners of y or y_0 be denoted by a_i, $i = 1, 2, \ldots, m$, where

$$a = a_0 < a_1 < a_2 < \cdots < a_m < a_{m+1} = b.$$

Suppose also that y_0 satisfies the Erdmann corner condition at its corners and the Euler equation elsewhere. We can now investigate the difference $J(y) - J(y_0)$:

$$J(y) - J(y_0) = \int_a^b [f(x, y(x), y'(x)) - f(x, y_0(x), y_0'(x))]\, dx.$$

From (3.15) we can write

$$J(y) - J(y_0) \geq \int_a^b [f_y^0(x)(y(x) - y_0(x)) + f_{y'}^0(x)(y'(x) - y_0'(x))]\, dx,$$

where f_y^0 and $f_{y'}^0$ have their usual meaning (cf. (2.4)). Integration by parts of the two terms leads to

$$\int_a^b f_{y'}^0(x)[y'(x) - y_0'(x)]\, dx = \sum_{i=1}^{m+1} [f_{y'}^0(x)(y(x) - y_0(x))]_{x=a_{i-1}}^{x=a_i} - \int_a^b \frac{d}{dx} f_{y'}^0(x)[y(x) - y_0(x)]\, dx.$$

The first term on the right side is zero, since at $x = a$ and at $x = b$ the functions y and y_0 agree and since y_0 satisfies the first Erdmann corner condition.

In summary, therefore,

$$J(y) - J(y_0) \geq \int_a^b \left[f_y^0(x) - \frac{d}{dx} f_{y'}^0(x) \right] (y(x) - y_0(x))\, dx = 0,$$

since y_0 was assumed to be an extremal of the problem. This concludes the proof of the following theorem.

(3.16) Theorem

If the integrand f of the variational problem is convex (cf. (3.13), (3.14)) for each x and with respect to the variables (y, y'), then *the following holds: A piecewise smooth, admissible function y_0, which at its corners satisfies the Erdmann corner conditions and which satisfies the Euler equation elsewhere, is a solution of the variational problem.*

Example
For the problem of the shortest curve joining two points A and B in the plane, the integrand $f(x, y, y') = \sqrt{1 + y'^2(x)}$ is convex since

$$f_{yy} = f_{yy'} = 0, \quad f_{y'y'} = (1 + y'^2(x))^{-3/2} > 0$$

and

$$f_{yy} f_{y'y'} - f_{yy'}^2 = 0.$$

As a consequence, the line segment joining A and B, since it is an admissible function and a uniquely determined extremal of the problem, is a solution of the problem.

Exercises
(3.17) Among all smooth functions $y : [0, 1] \mapsto \mathbb{R}$ with $y_0 = 0$ and $y_1 = 2$, find those functions for which the integral

$$\int_0^1 y^2(x)\, y'^2(x)\, dx$$

has its minimal value. Determine also whether there are piecewise smooth functions y with $y(0) = 0$ and $y(1) = 2$ but which are not extremals.

(3.18) What is the solution of the Ramsey model

$$\int_{t_0}^{t_1} a(bK(t) - K'(t) - C^*)^2\, dt \rightarrow \min, \quad a, \ b > 0,$$

if given boundary values are $K(t_0) = K_0 > 0$ and $K(t_1) = K_1$?

(3.19) Show that Theorem (3.16) cannot be applied to examples (1.2) and (1.4). What is the situation with respect to Example (1.3)? With respect to the problem of the shortest curve joining points A and B on a sphere?

4

The necessary conditions of Weierstrass and Legendre

If the integrand f of the variational problem is not convex, then we must seek other methods of solution. Clearly, each strong or weak solution y_0 satisfies certain conditions which relate to the individual points $(x, y_0(x))$ of the graph of y_0.

4.1 THE WEIERSTRASS NECESSARY CONDITION

The *Weierstrass necessary condition* is a condition for a strong solution y_0. To derive it we compare the variational integral $J(y_0)$ with the variational integral of functions belonging to a one-parameter family of neighbouring curves which, of course, must be admissible. Since these neighbouring functions may have corners, it is appropriate that we have weakened the original conditions for admissibility.

We will assume in this chapter, therefore, that y_0 is a smooth strong solution of the variational problem with the weakened conditions for admissibility. Thus, according to (2.30), y_0 is an extremum of the problem and so satisfies the Euler equation. Suppose now that x_0 is an arbitrary but fixed value with $a < x_0 \leq b$ and that $y_1 : [a, x_0] \mapsto \mathbb{R}$ is an arbitrary, smooth function

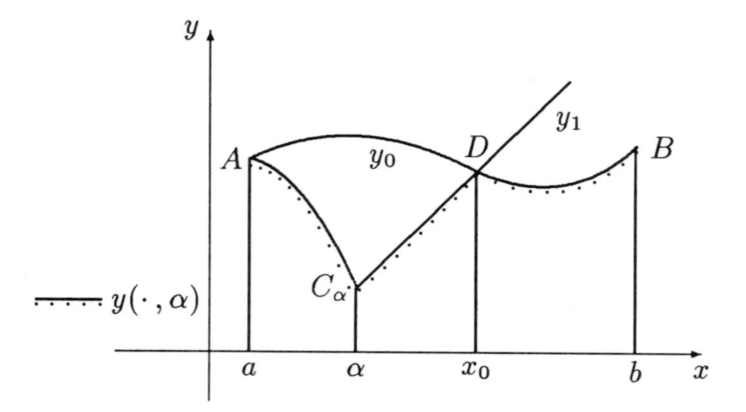

Fig. 4.1. Necessary condition of Weierstrass.

with the property $y_1(x_0) = y_0(x_0)$. Thus, y_1 can be a linear function $y_1(x) = c_1x + c_2$, $c_1, c_2 \in \mathbb{R}$, as shown in Fig. 4.1.

We define now a family $y(\cdot, \alpha)$ of functions with the parameter α, $a < \alpha \leq x_0$,

$$y(x, \alpha) = \begin{cases} y_2(x, \alpha) & \textit{if} \quad a \leq x \leq \alpha \leq x_0, \\ y_1(x) & \textit{if} \quad \alpha \leq x \leq x_0, \\ y_0(x) & \textit{if} \quad x_0 \leq x \leq b, \end{cases} \tag{4.1}$$

where

$$y_2(x, \alpha) = y_0(x) + \frac{x - a}{\alpha - a}(y_1(\alpha) - y_0(\alpha)).$$

It is easy to show that each function in the family satisfies the widened admissibility conditions but, in general, is no longer smooth.

The function y_0 belongs to the family $y(\cdot, \alpha)$ and corresponds to the parametric value $\alpha = x_0$, which is the right endpoint of the α-interval. Since $y(x, \alpha)$ is a continuous function of α, we can say that

$$d_0(y(\cdot, \alpha), y_0) < \varepsilon$$

when $\mid x_0 - \alpha \mid$ is sufficiently small. Consequently,

$$\hat{\Phi}(\alpha) = J(y(\cdot, \alpha)) - J(y_0) \geq 0 = \hat{\Phi}(x_0) \tag{4.2}$$

when $|x_0 - \alpha|$ is small enough. It follows that the function $\hat{\Phi}$ has a relative minimum at the right boundary point of its domain of definition $[a, x_0]$.

If at the point $\alpha = x_0$ the function $\hat{\Phi}$ has a (one-sided) derivative $\hat{\Phi}'(x_0)$, then $\hat{\Phi}'(x_0) \leq 0$ must hold (cf. p. 15). In order to determine $\hat{\Phi}$ we will employ an abbreviated notation whereby the curves along which we integrate will be denoted by their endpoints (cf. Fig. 4.1). Thus,

$$\begin{aligned}\hat{\Phi}(\alpha) &= \int_{AC_\alpha} f\,dx + \int_{C_\alpha D} f\,dx + \int_{DB} f\,dx - \int_{AB} f\,dx \\ &= \int_{AC_\alpha} f\,dx + \int_{C_\alpha D} f\,dx - \int_{AD} f\,dx.\end{aligned}$$

We now show that $\hat{\Phi}$ is in fact differentiable at the point $\alpha = x_0$

$$\begin{aligned}\hat{\Phi}'(x_0) &= \frac{d}{d\alpha}\left[\int_a^\alpha f\left(x, y_2(x,\alpha), \frac{\partial}{\partial x}y_2(x,\alpha)\right)\,dx\right. \\ &\quad \left.+ \int_\alpha^{x_0} f(x, y_1(x), y_1'(x))dx\right]_{\alpha = x_0} \\ &= f\left(x_0, y_0(x_0), \frac{\partial}{\partial x}y_2(x_0, x_0)\right) + \int_a^{x_0}\left[f_y \frac{\partial}{\partial\alpha}y_2(x, x_0)\right. \\ &\quad \left.+ f_{y'}\frac{\partial}{\partial\alpha}\frac{\partial}{\partial x}y_2(x, x_0)\right]\,dx - f(x_0, y_1(x_0), y_1'(x_0)).\end{aligned}$$

Here the arguments for the abbreviated expressions f_y and $f_{y'}$ are x, $y_2(x, x_0)$, and $\frac{\partial}{\partial x}y_2(x, x_0)$. Since the mixed second partials of y_2 are continuous, it holds that

$$\frac{\partial}{\partial\alpha}\frac{\partial}{\partial x}y_2(x,\alpha) = \frac{\partial}{\partial x}\frac{\partial}{\partial\alpha}y_2(x,\alpha).$$

By partial integration we obtain

$$\begin{aligned}&\int_a^{x_0} f_{y'}\frac{\partial}{\partial\alpha}\frac{\partial}{\partial x}y_2(x, x_0)\,dx = \left[f_{y'}\frac{\partial}{\partial\alpha}y_2(x, x_0)\right]_{x=a}^{x=x_0} \\ &- \int_a^{x_0}\frac{d}{dx}f_{y'}\left(\cdot, y_2(\cdot, x_0), \frac{\partial}{\partial x}y_2(\cdot, x_0)\right)(x)\frac{\partial}{\partial\alpha}y_2(x, x_0)\,dx.\end{aligned}$$

But from the definitions of y_1 and y_2:

(i) $y_1(x_0) = y_0(x_0)$ so that $y_2(x, x_0) = y_0(x)$ for all x and $\frac{\partial}{\partial x} y_2(x_0, x_0) =' y_0'(x_0)$;

(ii) $y_2(a, \alpha) = y_0(a)$ for all α so that $\frac{\partial}{\partial \alpha} y_2(a, \alpha) = 0$ for $\alpha > a$;

(iii) $y_2(\alpha, \alpha) = y_1(\alpha)$ for all α.

If we differentiate this last relation, we obtain

$$\frac{\partial}{\partial x} y_2(\alpha, \alpha) + \frac{\partial}{\partial \alpha} y_2(\alpha, \alpha) = y_1'(\alpha),$$

which, if substituted into the formula for $\hat{\Phi}(x_0)$, leads to

$$\begin{aligned} \hat{\Phi}'(x_0) =& f(x_0, y_0(x_0), y_0'(x_0)) \\ & - f(x_0, y_0(x_0), y_1'(x_0) + f_{y'}^0(x_0)[y_1'(x_0) - y_0'(x_0)] \\ & + \int_a^{x_0} \left[f_y^0(x) - \frac{d}{dx} f_{y'}^0(x) \right] \frac{\partial}{\partial \alpha} y_2(x, x_0)\, dx. \end{aligned} \tag{4.3}$$

(Cf. (2.4) for the definition of the abbreviations f_y^0 and $f_{y'}^0$.) The integral on the right side of (4.3) is equal to zero since y_0 is an extremal of the problem.

(4.4) Definition
The function $\mathcal{E} : G \times \mathbb{R} \times \mathbb{R} \mapsto \mathbb{R}$,

$$\mathcal{E}(x, y, p, q) = f(x, y, q) - f(x, y, p) - f_{y'}(x, y, p)(q - p),$$

is called the *Weierstrass $\mathcal{E}$-function (excess function).*

(4.5) Theorem (Weierstrass necessary condition)
Let y_0 be a strong solution of the variational problem with the weakened admissibility conditions (2.22). *If at a point $x \in [a, b]$ y_0 is smooth, then for all $q \in \mathbb{R}$*

$$\mathcal{E}(x, y_0(x), y_0'(x), q) \geq 0. \tag{4.6}$$

The proof will be carried out in two steps. In the first step we assume that y_0 is smooth on $[a, b]$. Then the function $\hat{\Phi}$ defined in (4.2) has a continuous derivative at the point x_0. Since $\hat{\Phi}$ has a relative minimum at the right boundary point of its domain of definition (cf. (4.2)), it must hold that $\hat{\Phi}(x_0) < 0$. Then from (4.3) it follows that

$$\hat{\Phi}'(x_0) = -\mathcal{E}(x_0, y_0(x_0), y_0'(x_0), y_1'(x_0)) \leq 0. \tag{4.7}$$

The slope $q = y_1'(x_0)$ of y_1 at the point x_0 may be chosen arbitrarily. Also, x_0 is an arbitrary point in the interval $(a, b]$.

For the second part of the proof we assume that y_0 is only piecewise smooth. Then, of course, y_0 is smooth on $(x_1, x_2) \subset [a, b]$ and $y_0 : [x_1, x_2] \mapsto \mathbb{R}$ is clearly a strong solution of the variational problem

$$J_{x_1 x_2}(y) = \int_{x_1}^{x_2} f(x, y(x), y'(x))\, dx \rightarrow \min,$$

with the boundary conditions $y(x_1) = y_0(x_1)$ and $y(x_2) = y_0(x_2)$. For otherwise y_0 could be varied on the interval $[x_1, x_2]$ to obtain a smaller value for the original variational integral J, in contradiction to the assumption that y_0 is a strong minimum. Thus, in accordance with the first part of the proof, y_0 satisfies the necessary Weierstrass condition for the variational problem with the integral $J_{x_1 x_2}$ at all points of (x_1, x_2). The excess function $\mathcal{E}$ does not depend, however, on the limits of the variational integral. So (4.6) follows, and the proof is complete.

1. Note
For reasons of continuity, it follows also that at the corners $x = a_k$ of y_0 and at $x = a$

$$\mathcal{E}(a_k, y_0(a_k), y_{0l}'(a_k), q) \geq 0, \quad \mathcal{E}(a_k, y_0(a_k), y_{0r}'(a_k), q) \geq 0.$$

2. Note
The meaning of the inequality $\mathcal{E}(x, y_0(x), y_0'(x), q) \geq 0$ is clearly seen in Fig. 4.2. For a fixed x we consider the function F of the

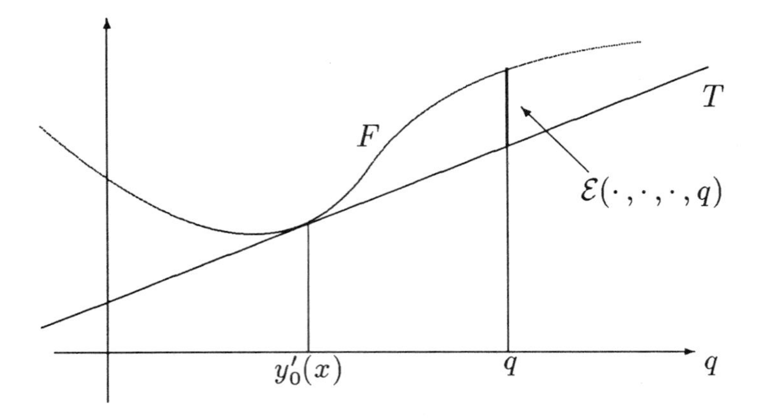

Fig. 4.2. The excess function of Weierstrass.

variable $q \in \mathbb{R} : F(q) = f(x, y_0(x), q)$. The function T defined by

$$T(q) = f(x, y_0(x), y_0'(x)) + f_{y'}(x, y_0(x), y_0'(x))(q - y_0'(x))$$

is, for fixed x, a linear function of q. Its graph is tangent to the graph of F at the point $q = y_0'(x)$. So considered, $\mathcal{E}(x, y_0(x), y_0'(x), q) = F(q) - T(q)$, and the Weierstrass necessary condition is satisfied when $F(q) \geq T(q)$ for all $q \in \mathbb{R}$. If, in addition, $\mathcal{E}(x, y_0(x), p, q) \geq 0$ for all p and q, then the graph of F lies everywhere above its tangent so that F is convex, as the theory of convex functions shows. This means, furthermore, that f regarded as a function of y' (x and y held fixed) is also convex. We will return to a consideration of the convexity of f considered as a function of y' in Example (9.22).

4.2 THE LEGENDRE NECESSARY CONDITION

The *Legendre necessary condition* is a condition for weak solutions. It is, in general, very easy to check. It may be regarded as a weak form of the Weierstrass condition. In order to prove Legendre's condition we will modify the derivation of the Weierstrass condition.

In what follows, y_0 will be a smooth, weak solution of the variational problem with broadened admissibility conditions. Then there exists a value $\delta > 0$ such that $J(y) - J(y_0) \geq 0$ for all admissible functions y whose d_1-distance from Y_0 is smaller than δ. We formulate this important technical point in a lemma.

(4.8) Lemma

Suppose

$$\begin{aligned} y_1(x) &= y_0(x_0) + (x - x_0)q, \\ y_2(x, \alpha) &= y_0(x) + \frac{x - a}{\alpha - a}(y_1(\alpha) - y_0(\alpha)) \quad (cf.\ (4.1)), \\ y(x, \alpha) &= \begin{cases} y_2(x, \alpha) & \text{if } a \leq x \leq \alpha \leq x_0, \\ y_1(x) & \text{if } \alpha \leq x \leq x_0, \\ y_0(x) & \text{if } x_0 \leq x \leq b. \end{cases} \end{aligned}$$

Then for each $\delta > 0$ there exists a number $\varepsilon > 0$ such that $d_1(y(\cdot, \alpha), y_0) < \delta$ when $\mid q - y_0'(x_0) \mid < \varepsilon$ and $\mid \alpha - x_0 \mid < \varepsilon$.

In order to prove the lemma an estimate of the distance $d_1(y(\cdot, \alpha), y_0)$ is obtained. The calculations are not difficult and so will not be included here. We make direct application of the lemma in what follows. The difference

$$\hat{\Phi}(\alpha) = J(y(\cdot, \alpha)) - J(y_0), \quad x_0 - \varepsilon < \alpha \leq x_0,$$

has a relative minimum at the point $\alpha = x_0$ if $d_1(y(\cdot, \alpha), y_0) < \delta$, or if $|q - y_0'(x_0)| < \varepsilon$. Thus

$$\begin{aligned} \hat{\Phi}'(x_0) = -\mathcal{E}(x_0, y_0(x_0), y_0'(x_0), q) \leq 0 \\ \text{if} \quad |q - y_0'(x_0)| < \varepsilon. \end{aligned} \tag{4.9}$$

A necessary condition stated in this form is not desirable, since the values δ and ε are in general not known explicitly.

Since, however, we have regarded the excess function $\mathcal{E}$ as a function of q, we obtain a satisfactory condition.

According to the definition of $\mathcal{E}$ (cf. (4.4)), $\mathcal{E}(x, y, p, p) = 0$.

$$\frac{\partial \mathcal{E}}{\partial q}(x, y, p, q) = f_{y'}(x, y, q) - f_{y'}(x, y, p),$$
$$\frac{\partial^2 \mathcal{E}}{\partial q^2}(x, y, p, q) = f_{y'y'}(x, y, q).$$

The Taylor expansion of $\mathcal{E}$ about $q = y_0'(x_0)$ gives us

$$\mathcal{E}(x_0, y_0(x_0), y_0'(x_0), q) = \tfrac{1}{2} f_{y'y'}(x_0, y_0(x_0), \tilde{q})(q - y_0'(x_0))^2,$$

where $\tilde{q}$ is a value between q and $y_0'(x_0)$. Since $f_{y'y'}$ is a continuous function and since q can be made arbitrarily close to $y_0'(x_0)$, it follows from (4.9) that $f_{y'y'}(x_0, y_0(x_0), y_0'(x_0)) \geq 0$. We summarize these results in a theorem.

(4.10) Theorem (Legendre necessary condition)
Let y_0 be a smooth, weak solution of the variational problem. Then for all $x \in [a, b]$

$$f_{y'y'}(x, y_0(x), y_0'(x)) \geq 0. \tag{4.11}$$

4.3 EXAMPLES AND PROBLEMS

We will illustrate the Weierstrass excess function for the class of variational integrals

$$\int_a^b g(x, y(x))\sqrt{1 + y'^2(x)}\, dx;$$

here $g(x, y)$ is an arbitrary function having continuous first, second and third derivatives. Recall here that $\sqrt{1 + y'^2(x)}\, dx = ds$, the element of arc length of the curve $x \mapsto (x, y(x))$ (cf. (1.5), (1.6), (1.8) and Exercise (2.17)). According to the definition of $\mathcal{E}$ we have

$$\mathcal{E}(x,y,p,q) = g(x,y)\left[\sqrt{1+q^2} - \sqrt{1+p^2} - \frac{p(q-p)}{\sqrt{1+p^2}}\right].$$

If we multiply this equation with $\sqrt{1+p^2}$, we obtain on the right-hand side

$$g(x,y)[\sqrt{(1+q^2)(1+p^2)} - (1+pq)].$$

Because $(p-q)^2 \geq 0$ it follows easily that

$$(1+q^2)(1+p^2) \geq (1+pq)^2,$$

so that for all $(x,y), \mathcal{E}(x,y,p,q) \geq 0$ when $g(x,y) \geq 0$.

From the above we infer that all of the extremals of the following variational problems satisfy the Weierstrass necessary condition: Problem (1.2) (smallest surface of revolution), Problem (1.3) (the Bernoulli problem), and Problem (1.1) (shortest curve joining two points in the plane). Recalling the necessary condition of Legendre, we note that for these problems

$$f_{y'y'}(x,y,y') = \frac{g(x,y)}{\sqrt{1+y^2}}.$$

For the next example we investigate the variational integral

$$\int_a^b (y'^2(x) - y^2(x))y'^4(x)\,dx.$$

The integrand does not depend explicitly on x; thus the extremals must satisfy

$$y'f_{y'}(x,y,y') - f(x,y,y') = c$$

so that

$$y'^2(x) - 3y^2(x)y'^4(x) = c \quad \text{for all} \quad x.$$

Here c is a constant of integration. If $c = 0$, then the extremals must be constant functions $y(x) = c^*$. If we check the Weierstrass and Legendre conditions for these extremals, we see that

$$\mathcal{E}(x, y, 0, q) = q^2 - y^2 q^4.$$

The extremal $y(x) = 0$ satisfies the Weierstrass necessary condition, since $\mathcal{E}(x, 0, 0, q) = q^2 \geq 0$. Contrarily, for the function $y(x) = c^* \neq 0$, the Weierstrass function as a function of q can take on both positive and negative values. Thus the extremal $y(x) = c^* \neq 0$ cannot be a strong solution of the variational problem since it does not satisfy the Weierstrass necessary condition. Whether these functions can be candidates for weak solutions, however, remains to be discussed, since they do satisfy the Legendre necessary condition:

$$f_{y'y'}(x, y, y') = 2 - 12y^2 y'^2, \qquad f_{y'y'}(x, c^*, 0) = 2 > 0.$$

(4.12) Exercises

For the variational problems having the following integrands, find the corresponding Weierstrass excess function and check whether $\mathcal{E}(x, y, p, q) \geq 0$ on the appropriate domain of definition for $\mathcal{E}$. Also, find $f_{y'y'}(x, y, y')$.

(a) $f(x, y, y') = y\sqrt{1 - y'^2}$ (cf. problem of Dido (1.4)).

(b) $f(x, y, y') = \sqrt{\cos^2 y + y'^2}$ (problem of the shortest arc joining two points A and B on the sphere (cf. (2.14)).

(c) $f(x, y, y') = (by - y' - c)^2$, where b and c are constants; $b,\ c > 0$ (cf. Ramsey growth model (2.13)).

5

The necessary condition of Jacobi

We recall the investigation of the extrema of a differentiable function Φ. The condition $\Phi'(x_0) = 0$, which is necessary for a relative minimum (and which led us in Chapter 2 to the Euler equation), is also satisfied if x_0 provides a relative maximum, an inflection point or, generally, a stationary point for Φ. It is, therefore, not a specific condition for a relative minimum. The Taylor formula for Φ at the point x_0 yields

$$\Phi(x) = \Phi(x_0) = \Phi'(x_0)(x - x_0) + \tfrac{1}{2}\Phi''(x^*)(x - x_0)^2,$$

where x^* is an intermediate value between x_0 and x. It follows from this that, if $\Phi'(x_0) = 0$, then x_0 yields a relative minimum provided

$$\Phi(x) - \Phi(x_0) = \tfrac{1}{2}\Phi''(x^*)(x - x_0)^2 \geq 0$$

for all values x in a sufficiently small ε-neighbourhood of x_0. If Φ'' is continuous, then x_0 can at most be a relative minimum of Φ if

$$\Phi''(x_0) \geq 0. \tag{5.1}$$

By analogy, various consequences can be derived for the variational problem $J(y) \to \min$. In this way we obtain a new proof for the necessary Legendre conditions. But we can also derive the necessary condition of Jacobi, which is a condition on the entire solution curve rather than about single points or small pieces of the solution. Thus, for example, for the variational problem of the shortest curve joining two points on the sphere, this condition requires that the solutions of the problem be found only among pieces of great circles which are not longer than half the circumference of the sphere (cf. Example 2 in Section 5.2).

5.1 THE SECOND VARIATION AND THE ACCESSORY PROBLEM

In this section we will investigate the expression $\Phi''(0)$ belonging to the functional $\Phi(\alpha) = J(y_\alpha)$ as defined in Chapter 2, and whereby $y_\alpha(x) = y_0(x) + \alpha Y(x)$. Since $\frac{1}{2}\alpha^2\Phi''(0)$ is the element of the second order in the Taylor expansion of Φ about the point $\alpha = 0$, $\Phi''(0)$ is called the *second variation* of the functional. $\Phi''(0)$ is an integral which may be represented as a functional. This new variational problem is called the *accessory variational problem* and is generally of simpler structure than the initial problem.

So we suppose that y_0 is a strong solution or a weak solution of the variational problem $J(y) \to \min$ with the weakened boundary conditions (2.22). We form the family of admissible functions $y_\alpha = y_0 + \alpha Y$ from y_0 and a piecewise differentiable function $Y : [a, b] \mapsto \mathbb{R}$, which satisfies the boundary conditions $Y(a) = Y(b) = 0$ but otherwise is arbitrary. Since y_0 is a strong (or weak) solution, there is a positive number δ such that $J(y) - J(y_0) \geq 0$ holds for all admissible functions y such that the distance d_0 (or d_1) is less than δ. We choose $\alpha_0 > 0$ so that the distance $\| y_\alpha - y_0 \|_1 < \delta$ for all α such that $|\alpha| < \alpha_0$. Then if the function $\Phi(\alpha) = J(y_\alpha)$ (whose derivative is smooth) has a relative minimum at $\alpha = 0$, it must follow that $\Phi''(0) \geq 0$. Thus we consider here

$$\Phi''(0) = \int_a^b [f^0_{yy}(x)Y^2(x) + 2f^0_{yy'}(x)Y(x)Y'(x) + f^0_{y'y'}(x)Y'^2(x)]\,dx = I(Y), \tag{5.2}$$

where we have the abbreviations

$$\begin{aligned} f^0_{yy}(x) &= f_{yy}(x, y_0(x), y_0'(x)), \\ f^0_{yy'}(x) &= f_{yy'}(x, y_0(x), y_0'(x)), \\ f^0_{y'y'}(x) &= f_{y'y'}(x, y_0(x), y_0'(x)). \end{aligned} \tag{5.3}$$

The integral (5.2) is the *accessory variational integral* I(y) and has the integrand

$$2\Omega(x, Y, Y') = f^0_{yy}(x)Y^2 + 2f^0_{yy'}(x)YY' + f^0_{y'y'}(x)Y'^2. \tag{5.4}$$

Here Y and Y' denote variables. Admissible functions for the accessory variational integral problem are all piecewise smooth functions $Y : [a, b] \mapsto \mathbb{R}$ which satisfy the boundary conditions $Y(a) = Y(b) = 0$. The null function $Y_0 = 0$ is an admissible function. Clearly, $I(Y_0) = 0$. Thus, the necessary condition deriving from the requirement that $\Phi''(0) = I(Y) \geq 0$ may be stated as follows: The minimal value of the accessory variational integral $I(Y)$ must be zero. We want to bring these conditions into a form which can be applied more easily. So we need to consider the accessory variational problem.

One of the difficulties we face consists of this: The integrand 2Ω does not in general possess continuous first, second and third partial derivatives, as was required of the integrands of the variational problem of Section 1.2. We can circumvent this difficulty if we assume higher derivatives of f and y_0, the functions from which Ω is formed. Thus, Ω will have continuous first, second and third partial derivatives, if we assume that f has continuous fifth derivatives and y_0 has continuous fourth derivatives. These additional assumptions mean, of course, that we must exclude certain variational problems from further investigation. A careful analysis shows, however, that these additional assumptions are not really needed.

The Euler equation for the extremals $Y : [a,b] \mapsto \mathbb{R}$ of the accessory variational problem is

$$\begin{aligned}\frac{d}{dx}[\Omega_{Y'}(\cdot, Y, Y')](x) &- \Omega_Y(x, Y(x), Y'(x)) \\ &= \frac{d}{dx}[f^0_{yy'}Y + f^0_{y'y'}Y'](x) \\ &- f^0_{yy}Y(x) - f^0_{yy'}Y'(x) = 0.\end{aligned} \tag{5.5}$$

Equation (5.5) is called the *Jacobi (differential) equation.* If the extremal Y is twice differentiable, then one can carry out the differentiation in (5.5). In the event that $f^0_{y'y'}(x) \neq 0$, the Jacobi equation is locally solvable for Y'' and so is a homogeneous linear differential equation of second order for the function Y.

5.2 JACOBI'S NECESSARY CONDITION

Jacobi's necessary condition is an important consequence of the condition that, for either a strong or weak solution y_0 of the variational problem $J(y) \mapsto \min$, the minimal value of the accessory variational integral is zero. We come upon the Jacobi condition when we investigate the question whether the accessory problem can have solutions other than the null solution $Y = 0$. A result of this investigation is that for 'good' (that is, in the sense of (5.9), regular) solutions y_0, only the null solution is possible.

We will need the following as a preliminary to the proof of *Jacobi's necessary condition.*

(5.6) Theorem
If on $[a,c]$ the smooth function Y is a solution of the Jacobi equation (5.5) which satisfies the boundary conditions $Y(a) = Y(c) = 0$, where $c \leq b$, then

$$\int_a^c \Omega(x, Y(x), Y'(x))\, dx = 0.$$

Proof
From the definition (5.4) of Ω we obtain the relation

$$2\Omega(x, Y, Y') = \Omega_Y(x, Y, Y')Y + \Omega_{Y'}(x, Y, Y')Y' \text{ for } Y,\ Y' \in \mathbb{R}.$$

Using this representation of the integrand and employing integration by parts, we obtain

$$\int_a^c 2\Omega(x, Y(x), Y'(x))\, dx [\Omega_{Y'}(x, Y(x), Y'(x))Y(x)]_{x=a}^{x=c}$$
$$+ \int_a^c \left[\Omega_Y(x, Y(x), Y'(x)) - \frac{d}{dx}\Omega_{Y'}(\cdot, Y, Y')(x)\right] Y(x)\, dx.$$

The first term on the right vanishes because of the boundary values $Y(a) = Y(c) = 0$. The second vanishes because Y is a solution of the Jacobi equation. This concludes the proof.

In what follows we will be concerned essentially with the question whether the Jacobi equation has solutions, other than the null function, which possess the properties named in Theorem (5.6). In order to express these matters more compactly we introduce the concept of conjugate points.

(5.7) Definition
Let $y_0 : [a_0, b] \mapsto \mathbb{R}$ be an extremal for the variational problem $J(y) \to \min$ and suppose y_0 has continuous first, second, third and fourth derivatives. If there is a function $Y : [a_0, c] \mapsto \mathbb{R}$ which has continuous first and second derivatives, which is a solution of the Jacobi equation for y_0, which satisfies the boundary conditions $Y(a_0) = Y(c) = 0$, and which on each subinterval of positive length of its domain of definition is not identically zero, then c is called a *conjugate point* of a_0 for the extremal y_0.

We can now formulate Jacobi's necessary condition:

(5.8) Theorem
Let $y_0 : [a, b] \mapsto \mathbb{R}$ have continuous first, second, third and fourth derivatives, and suppose y is a weak solution of the variational

problem. The point $c \in (a,b)$ can only be a conjugate point of $x = a$ if

$$f^0_{y'y'}(c) = f_{y'y'}(c, y_0(c), y_0'(c)) = 0.$$

Note: We have already encountered the function $f^0_{y'y'}$ at several points. The necessary condition of Legendre requires $f^0_{y'y'}(x) \geq 0$. For the Euler and Jacobi differential equations $f^0_{y'y'} \neq 0$ is the condition for regularity. If it is satisfied, then the equations can be solved, respectively, for y'' and Y'', the highest derivatives of the functions sought. On the basis of this we make the following definition.

(5.9) Definition

The triple (x, y, y') is called *regular* if

$$f_{y'y'}(x, y, y') \neq 0$$

and *singular* if $f_{y'y'}(x, y, y') = 0$. A differentiable function y is said to be *regular* if, for all x values in the domain of definition,

$$f_{y'y'}(x, y(x), y'(x)) \neq 0.$$

Before we prove Jacobi's necessary condition, we provide some examples of extremals which satisfy or do not satisfy Jacobi's condition.

Example 1 Consider the variational problem for the shortest curve joining two points in the plane:

$$J(y) = \int_a^b \sqrt{1 + y'^2(x)}\, dx \to \min.$$

Extremals are line segments $y(x) = cx + d$, where c and d are constants. The integrand $2\Omega(x, Y, Y')$ of the accessory problem (for $y_0(x) = c_0 x + d_0$) is

$$2\Omega(x, Y, Y') = f_{y'y'}(x, y_0(x), y_0'(x))Y'^2 = \tfrac{1}{2}c^* Y'^2,$$

where $c^* = 1/\sqrt{(1+c_0^2)^3}$. This leads us to the Jacobi equation

$$\frac{d}{dx}[c^*Y'] = c^*Y'' = 0.$$

The solutions of this equation are, therefore,

$$Y(x) = Cx + D,$$

where C and D are integration constants.

A solution of the Jacobi equation which is not the trivial solution ($\equiv 0$) has at most one zero. Thus, for the extremal y_0, there can be no point which is conjugate to $x = a$. All extremals of this problem satisfy Jacobi's necessary condition.

Example 2 Consider the shortest curve joining two points on the sphere (cf. Section 2.3):

$$J(y) = \int_a^b \sqrt{\cos^2 y(x) + y'^2(x)}\, dx \to \min.$$

Here x is the longitude and $y(x)$ is the latitude of the point on the curve. The equator $y_0(x) = 0$ is an extremal; it is regular since $f_{y'y'}(x, 0, 0) = 1 \neq 0$. The integrand $2\Omega(x, Y, Y')$ of the accessory problem for $y_0 = 0$ is

$$2\Omega(x, Y, Y') = -Y^2 + Y'^2.$$

Thus, the Jacobi equation

$$Y'' + Y = 0$$

has solutions of the form $Y(x) = C \cos x + D \sin x$, where C and D are constants of integration.

Now we will check whether for $y_0 = 0$ there is a point conjugate to $x = 0$. To do this, we seek a solution Y of the Jacobi equation which has the initial value $Y(0) = 0$. Clearly, the function is $Y(x) = D \sin x$. If $D \neq 0$, then $Y(x) = D \sin x$ has a zero at the point $x = k\pi$, where k is an integer. Thus, $x = \pi$ is a conjugate point for $x = 0$ for the extremal y_0, and in the open interval $(0, \pi)$ there is no point conjugate to $x = 0$ for y_0. In the event that the domain of definition $[0, b]$ of y_0 lies in $[0, \pi]$, then Jacobi's necessary condition for y_0 is satisfied; otherwise, it is not. Thus, with Theorem (5.8), we can conclude that the extremal $y_0 : [0, b] \mapsto \mathbb{R}, y_0(x) = 0, b > \pi$, is not a weak solution of the variational problem and so, *a fortiori*, is no solution at all.

If $b \geq \pi$, then y_0 with a domain of definition $[0, b]$ represents a part of the equator that is longer than one- half of the equator. This piece of the equator is clearly not the shortest curve joining its endpoints.

From this example we can see that there are extremals whose 'shorter' parts are solutions but whose 'longer' parts are not. At issue here is the domain of definition of the extremal.

Proof of Theorem (5.8)

To begin, we consider a point $c \in (a, b)$ which for the extremal y_0 is conjugate to the point $x = a$. From definition (5.7) there is a solution Y_0 of the Jacobi equation which has continuous first and second derivatives, which satisfies the conditions $Y(a) = Y(c) = 0$, and which is not identically zero on any interval of positive length. Now let the function $Y^* : [a, b] \mapsto \mathbb{R}$ be defined as follows:

$$Y^*(x) = \begin{cases} Y_0(x) & \text{if} \quad x \in [a, c), \\ 0 & \text{if} \quad x \in [c, b]. \end{cases}$$

Y^* is piecewise smooth and is an admissible function of the accessory variational problem. On the interval (a, c), Y^* satisfies the Jacobi equation. Thus we may apply Theorem (5.6) to obtain

$$I(Y^*) = \int_a^c 2\Omega(x, Y_0(x), Y_0'(x))\, dx + \int_c^b 2\Omega(x, 0, 0)\, dx = 0.$$

Since y_0 is a strong or weak solution of the variational problem, it follows the accessory variational integral $I(Y) \geq 0$. Y^* is thus a solution of the accessory variational problem. From the theory of the first variation applied to the accessory problem it follows that:

(i) either Y^* has a corner at the point $x = c$,

or

(ii) Y^* is smooth at $x = c$.

We will discuss these two possible cases separately. In case (i) the right- and left-hand derivatives $Y_r^{*\prime}(c)$ and $Y_l^{*\prime}(c)$ are unequal

$$Y_r^{*\prime}(c) = 0 \neq Y_l^{*\prime}(c) = Y_0'(c). \tag{5.10}$$

Y^* satisfies the Erdmann corner conditions at the point $x = c$

$$\Omega_{Y'}(c, Y^*(c), Y_l^{*\prime}(c)) = \Omega_{Y'}(c, Y^*(c), Y_r^{*\prime}(c)).$$

Since $Y^*(c) = 0$, we can infer from the corner condition that

$$\begin{aligned} \Omega_{Y'}(c, Y^*(c), Y_l^{*\prime}(c)) &= f^0_{y'y'}(c) Y_0'(c) \\ &= \Omega_{Y'}(c, Y^*(c), Y_r^{*\prime}(c)) = 0. \end{aligned}$$

Since according to (5.10) $Y_0'(c) \neq 0$, we must have $f^0_{y'y'}(c) = 0$, as claimed in Theorem (5.8).

For case (ii) we apply the proof indirectly. That is, contrary to the conclusion of Theorem (5.8), we assume that $f^0_{y'y'}(c) \neq 0$ and proceed to show that this assumption leads to a contradiction. If the value of the continuous function $f^0_{y'y'}$ is different from zero at the point $x = c$, then there must be a neighbourhood of $x = c$ in which $f^0_{y'y'}$ has no zeros so that on this interval the Jacobi equation can be solved for Y''. According to the existence and uniqueness theorems for initial value problems, there is one and only one solution Y of the Jacobi equation which has a smooth derivative, is defined on the neighbourhood of $x = c$, and which satisfies $Y(c) = 0$ and $Y'(c) = 0$. The null function is clearly this uniquely determined solution. On the other hand, Y^* also satisfies the initial conditions $Y^*(c) = 0$, $Y_l^{*\prime}(c) = Y_r^{*\prime}(c) = 0$. Y^* also has a smooth derivative (cf. Theorem (2.31) applied to the accessory problem and to the continuously differentiable extremal Y^*). Consequently, Y^* must be the null function in a neighbourhood of the point $x = c$, so that Y^* vanishes on a subinterval $[a, c]$ which has positive length. This assertion is a contradiction of the statement that $Y^*(x) = Y_0(x)$ for $x \in [a, c]$ and that Y_0 is not identically zero on any interval of positive length. Thus the claim that $f^0_{y'y'}(c) \neq 0$ is false and the proof is complete.

Note: One can formulate Jacobi's necessary condition in the following way: Let $y_0 : [a, b] \mapsto \mathbb{R}$ be a regular weak or strong solution of the variational problem. Then the interval $[a, b)$ contains no points conjugate to $x = a$.

We will use the argument of the proof above (for Theorem (5.8)) to show that if the extremal y_0 is regular and has a conjugate point $x = c$ with respect to the point a in the interval $[a, b)$, then the accessory integral I for y_0 has a negative value. So let the function Y^* be defined as in the proof above. Again, $I(Y^*) = 0$. But since y_0 is regular at the point $x = c$, Y^* must have a corner at $x = c$ even though it does not satisfy the Erdmann corner condition. It follows that Y^* is not a solution of the accessory equation and $I(Y^*) = 0$ is not the minimal value of I.

Exercises

(5.11) Consider the variational problem

$$J(y) = \int_0^1 (y'^2(x) - y^2(x))\, dx \to \min, \quad y(0) = y(b) = 0.$$

Show that the null function $y_0 \equiv 0$ is a regular extremal of the problem. Are there additional extremals that satisfy the boundary conditions?

Find values $b > 0$ for which the null function $y_0 \equiv 0$ is not a weak solution of the problem.

(5.12) For the problem (1.2) of the smallest surface of revolution

$$J(y) = \int_0^1 y(x)\sqrt{1 + y'^2(x)}\, dx \to \min; \; y(0) = 1, y(1) = b,$$

there are, for certain boundary values, two extremals which satisfy the boundary conditions. It may be presumed, therefore, that one of these extremals is not a solution of the variational problem. Thus, show that for the extremal $y(x, \alpha, \beta) = (1/\alpha)\cosh(\alpha x + \beta)$ the Jacobi equation has the form

$$Y'' - 2\alpha \tanh(\alpha x + \beta)\, Y' + \alpha^2 Y = 0,$$

and that the general solution of this differential equation is

$$Y(x) = \frac{C}{\alpha}\sinh(\alpha x + \beta) + D[\frac{x}{\alpha}\sinh(\alpha x + \beta) - \frac{1}{\alpha^2}\cosh(\alpha x + \beta)],$$

where C and D are real constants.

Show further that the extremal $y_0 : [0, b] \mapsto \mathbb{R}, \quad y_0(x) = \cosh x$, satisfies the necessary condition independently of how $b > 0$ is chosen.

Show that for the extremal $y : [0,1] \mapsto \mathbb{R}, \ y(x) = -(1/(2\beta)) \cosh(\beta - 2\beta x)$, where $\beta \in [-1, 0]$ is a solution of the equation $\cosh \beta + 2\beta = 0$, there is a point $c < 1$ which is conjugate to $x = 0$. Since y is a regular extremal, y cannot be a weak solution of the variational problem.

(5.13) Consider the variational problem

$$J(y) = \int_0^1 \sqrt{y(x)[1 + y'^2(x)]}\, dx \rightarrow \min$$

with boundary conditions $y(0) = 2$ and $y(1) = 5$. Find the extremals (of the problem) which satisfy the given boundary conditions. *Hint:*

$$y(x, \alpha, \beta) = \alpha + \frac{1}{\alpha}\left(\frac{x + \beta}{2}\right)^2$$

defines the two-parameter *family of extremals* where α and β are the parameters. The boundary conditions are satisfied for pairs $(\alpha, \beta) = (1, 2)$ and $(1/13, -10/13)$, respectively.

Check whether the extremals satisfy Jacobi's necessary condition. *Hint:* For the extremal for $\alpha = 1$ and $\beta = 2$,

$$Y(x, A_1, B_1) = A_1\left[1 - \left(\frac{x}{2} + 1\right)^2\right] + B_1(x + 2)$$

is the general solution of the Jacobi equation. Here A_1 and B_1 are parameters of the family of solutions. For the extremals with $\alpha = 1/13$ and $\beta = -10/13$, the general solution of the Jacobi equation is

$$Y(x, A_2, B_2) = A_2 \left[\frac{1}{13} - (\frac{13x - 10}{2})^2\right] + B_2(13x - 10).$$

Compare this problem with Problem (5.19).

5.3 ONE-PARAMETER FAMILIES OF EXTREMALS AND THE GEOMETRIC SIGNIFICANCE OF CONJUGATE POINTS

With the help of a family of extremals we want next to illustrate the concept of the conjugate point. To be sure, this geometric description of the conjugate point will be appropriate only in a few cases. But it may explain somewhat more clearly the role of the conjugate point than the definition and the Jacobi condition do. The two theorems about families of extremals which we prove here will also be needed in subsequent sections. They show the connection between families of extremals and the Jacobi equation.

(5.14) Theorem
Let $\tilde{y} : [a,b] \times [-\alpha^*, \alpha^*] \mapsto \mathbb{R}$, $(x, \alpha) \mapsto \tilde{y}(x, \alpha)$ *be a one-parameter family of extremals; i.e. a function with the following properties:*

(i) *For each fixed* $\alpha \in [-\alpha^*, \alpha^*]$ *there is a function* $\tilde{y}(\cdot, \alpha) : [a,b] \mapsto \mathbb{R}$, $x \mapsto \tilde{y}(x, \alpha)$, *which is an extremal of the variational integral* J.

(ii) $\tilde{y}$ *possesses continuous first, second, third and fourth derivatives with respect to* x *and at least first and second continuous derivatives with respect to* α. *Also the mixed derivatives of* $\tilde{y}$ *are continuous.*

Then $Y : [a, b] \mapsto \mathbb{R}$,

$$Y(x) = \tilde{y}_\alpha(x, 0) = \tfrac{\partial}{\partial\alpha}\tilde{y}(x, 0),$$

is a solution of the Jacobi equation for the extremal $y_0 = \tilde{y}(\cdot, 0)$.

Proof

The extremals $\tilde{y}(\cdot, \alpha)$ satisfy the Euler equation. Thus, for all α such that $|\alpha| < \alpha^*$,

$$\frac{d}{dx}[f_{y'}(\cdot, \tilde{y}(\cdot, \alpha), \tilde{y}_x(\cdot, \alpha))](x) - f_y(x, \tilde{y}(x, \alpha), \tilde{y}_x(x, \alpha)) = 0.$$

We differentiate this equation partially with respect to α at the point $\alpha = 0$. Since the second mixed derivatives of $\tilde{y}$ are continuous, the orders of differentiation may be interchanged. Then we obtain

$$\begin{aligned} 0 &= \frac{d}{dx}[f^0_{yy'}\tilde{y}_\alpha(\cdot, 0) + f^0_{y'y'}\tilde{y}_{x\alpha}(\cdot, 0)](x) \\ &\quad - f^0_{yy}(x)\tilde{y}_\alpha(x, 0) - f^0_{yy'}(x)\tilde{y}_{x\alpha}(x, 0) \\ &= \frac{d}{dx}[\Omega_{Y'}(\cdot, Y, Y')](x) - \Omega_Y(x, Y(x), Y'(x)), \end{aligned}$$

where 2Ω is the integrand of the accessory variational problem belonging to $y_0 = \tilde{y}(\cdot, 0)$ and where $\tilde{y}_\alpha(\cdot, 0)$ is abbreviated by Y. (For the abbreviations $f^0_{yy'}$ and the like, see (5.3).) Thus Y satisfies the Jacobi equation, which was to be shown.

Note: If we expand $\tilde{y}(x, \alpha)$ about the point α we obtain

$$\tilde{y}(x, \alpha) = \tilde{y}(x, 0) + \alpha\tilde{y}_\alpha(x, 0) + \text{ remainder},$$

so that $\alpha\tilde{y}_\alpha(x, 0) = \alpha Y(x)$ is an approximation of the difference $\tilde{y}(x, \alpha) - \tilde{y}(x, 0)$ of the two extremals.

Theorem (5.14) has a converse: With each solution Y of the Jacobi equation a certain family of extremals can be associated.

(5.15) Theorem

Let y_0 be a regular extremal of the variational problem. Then for each solution Y of the Jacobi equation for y_0, which has a smooth second derivative, there is a one-parameter family of extremals $\tilde{y}$ for which $Y(x) = \tilde{y}_\alpha(x, 0)$.

Proof

Since $f^0_{y'y'}(x) \neq 0$ for all $x \in [a, b]$, the Euler and Jacobi differential equations may be solved (locally) for y'' and Y'', respectively. The Jacobi equation is a homogeneous linear differential equation of second order for Y. As we know, the solutions of a linear homogeneous differential equation of second order form a two-dimensional vector space. Thus we can represent each solution Y as a linear combination of two basis solutions (vectors) Y_1 and Y_2

$$Y = c_1 Y_1 + c_2 Y_2, \quad c_1, \ c_2 \quad \text{constant.} \tag{5.16}$$

The totality of the solutions y of the initial value problem determined by the Euler equation and the initial values,

$$y(a) = \alpha_1, \quad y'(a) = \alpha_2,$$

forms a two-parameter family of extremals $y^*(\cdot, \alpha_1, \alpha_2)$. According to the theorems asserting the dependence of the solutions of an initial value problem on the initial values, the function y^* has a smooth fourth derivative with respect to x and has continuous partial derivatives with respect to α_1 and α_2. Furthermore, the mixed derivatives with respect to x and α_1 or α_2 exist and are continuous. The functions Y_1 and Y_2,

$$Y_1(x) = y^*_{\alpha_1}(x, \alpha^0_1, \alpha^0_2), \quad Y_2(x) = y^*_{\alpha_2}(x, \alpha^0_1, \alpha^0_2), \tag{5.17}$$

where $\alpha^0_1 = y_0(a)$ and $\alpha^0_2 = y'_0(a)$, are solutions of the Jacobi equation, are in the family and are linearly independent. From the differentiation of the initial values,

$$y^*(a, \alpha_1, \alpha_2) = \alpha_1, \quad y^*_x(a, \alpha_1, \alpha_2) = \alpha_2,$$

we obtain

$$Y_1(a) = 1, \quad Y_1'(a) = 0,$$
$$Y_2(a) = 0, \quad Y_2'(a) = 1.$$

For an arbitrary solution $Y = c_1 Y_1 + c_2 Y_2$ of the Jacobi equation we define the one-parameter family of extremals

$$\tilde{y}(x, \alpha) = y^*(x, c_1\alpha + \alpha_1^o, c_2\alpha + \alpha_2^o).$$

We note here that $Y = \tilde{y}_\alpha(\cdot, 0)$.

We are now able to show the geometrical significance of a point conjugate to $x = a$. We assume that y_0 is a regular extremal. According to definition, the point $x = c$ is a conjugate point for $x = a$ if there is a solution Y (with smooth derivative) of the Jacobi equation which has $x = a$ and $x = c$ as zeros, but which on any interval of positive length is not identically zero.

Now Y may be written as a linear combination of the basis vectors defined in (5.17): $Y = c_1 Y_1 + c_2 Y_2$. Since $Y(a) = 0 = c_1 Y_1(a) + c_2 Y_2(a) = c_1$, it follows that $Y = c_2 Y_2$. Since Y is not the null function, $c_2 \neq 0$. Thus the family of extremals

$$\tilde{y}(x, \alpha) = y^*(x, \alpha_1^o, c_2\alpha + \alpha_2^o)$$

belonging to Y consists of extremals whose graphs pass through the point (a, y_a). If $\tilde{y}_\alpha(x, 0) \neq 0$ for $x > a$, then the equation

$$y = \tilde{y}(x, \alpha) \tag{5.18}$$

may be solved near $\alpha = 0$ for $\alpha : \alpha = \hat{\alpha}(x, y)$.

The function $\hat{\alpha}$ has continuous partial derivatives with respect to x and y. Thus, through each point (x, y) near the graph of y_0 there passes exactly one curve of the family of extremals. If $Y(c) = \tilde{y}_\alpha(c, 0) = 0$, then the family of extremals no longer has this attractive property. It can happen, as in the example of the shortest curve joining two points on a sphere, that the extremals of the family pass not only through (a, y_a) but also through a point $(c, y_0(c))$, a diametrical point. It can also happen

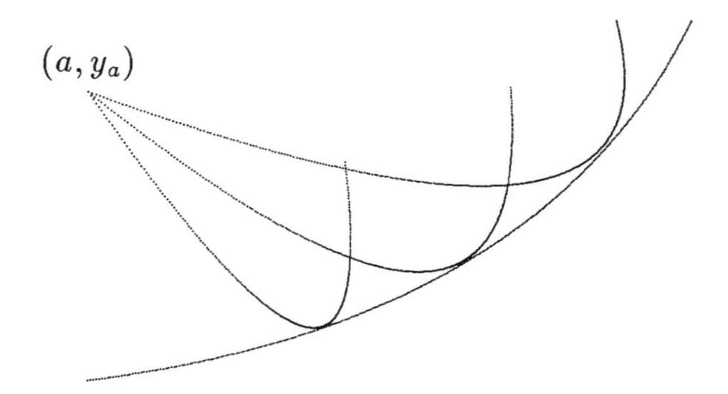

Fig. 5.1. Conjugate points for Jacobi's equation.

that the family of extremals has an envelope which is defined by the condition $\tilde{y}_\alpha(x, \alpha) = 0$ (cf. Fig. 5.1). Thus, the general case may be described by saying that while at $x = c$ there is an envelope for the family, it does not have to be a curve. Jacobi's necessary condition requires that y_0 be regular, that the curves of the family of extremals $\tilde{y}$ cover a neighbourhood U of the graph of $y_0 : (a, b) \mapsto \mathbb{R}$ in the xy-plane, and that the function $\hat{\alpha} : U \mapsto \mathbb{R}$, which is obtained from the solution of (5.18) for α, has continuous partial derivatives with respect to x and y.

(5.19) Problem

Consider the variational problem

$$J(y) = \int_0^1 \sqrt{y(x)(1 + y'^2(x))}\, dx \to \min$$

with the boundary conditions $y(0) = 2,\ y(1) = 5$. For the two extremals which satisfy the initial conditions, determine the general solution of the Jacobi equation (cf. Problem (5.13)). This problem gives an example showing that the solutions of the Jacobi equation may occasionally be found more easily with the help of the two-parameter family of extremals for the variational problem.

5.4 CONCLUDING REMARKS ABOUT NECESSARY CONDITIONS

At this point we summarize briefly what we have discovered thus far and how we normally proceed with solving the variational problem.

It is rewarding to check whether the integrand of the variational integral is convex in y and y'. If so, then one only has to determine the admissible functions which satisfy the Euler equation (2.8) and the Erdmann corner conditions (2.28) and (2.29). This done, the solution of the variational problem is complete.

If, however, the integrand f is not a convex function of y and y', then we proceed as follows:

(1) If only functions having continuous derivatives are admitted, then we solve the boundary value problem consisting of the Euler equation and the boundary conditions.

(1*) If, on the other hand, the weakened conditions for admissibility are accepted, then we seek all functions which satisfy the boundary conditions, which piecewise satisfy the Euler equation (2.8), and which satisfy the Erdmann corner conditions.

(2) We dispose of those functions which do not satisfy the Legendre necessary condition (4.10).

(2*) If a strong solution – or for that matter, any solution – of the variational problem is sought, then the Weierstrass necessary condition (4.6) must be satisfied.

(3) Finally, we check Jacobi's necessary condition (5.8) for those functions which satisfy the necessary conditions given above.

If by application of this selection process only one function y_0 remains, then it is tempting to regard this function y_0 as the solution of the problem. The following example shows that this assumption is, in general, false. How can this phenomenon be explained? It is explained by the fact that there are variational problems which simply have no solutions – and this is not always easily seen.

The following example shows that the satisfaction of all the necessary conditions may still not be sufficient for a strong solution.

(5.20) Example

Consider the variational problem whose integrand is

$$f(x, y, y') = y'^2 - y^2 y'^4,$$

and whose boundary conditions are

$$a = 0, \quad y_a = 0, \quad b = 1, \quad y_b = 0.$$

Since the integrand does not depend on x, the extremals must satisfy differential equation (2.12)

$$y'^2(x) - 3y^2(x)\, y'^4(x) = c_1 = \text{ constant.}$$

We will not determine the extremals here, but rather observe that $y_0 = 0$ is an extremal which satisfies the boundary conditions. Since

$$f_{y'y'}(x, 0, 0) = 2 \neq 0,$$

y_0 is regular. Also y_0 satisfies the Weierstrass necessary condition since

$$\mathcal{E}(x, 0, 0, q) = q^2 > 0 \quad \text{if} \quad q \neq y_0'(x) = 0.$$

Next we check Jacobi's necessary condition. For y_0 the integrand of the accessory variational problem is

$$2\Omega(x, Y, Y') = 2\,Y'^2.$$

A solution $Y \neq 0$ of the Jacobi equation $Y'' = 0$ has at most one zero. Thus for y_0 there is no point conjugate to $x = 0$ in $[0, 1]$. The function $y : [0, 1] \to \mathbb{R}$, defined by

$$y(x) = \begin{cases} \dfrac{cx}{d} & \text{if} \quad 0 \leq x \leq d, \\[2ex] \dfrac{c(1-x)}{1-d} & \text{if} \quad d < x \leq 1, \end{cases}$$

where c and d are constants and $1/2 < d < 1$, is an admissible function satisfying the weakened conditions (2.22). Also,

$$d_0(y, y_0) = \max_{x \in [1,0]} | y(x) - y_0(x) | = | c |$$

and

$$J(y) = \frac{c^2}{d} - \frac{c^6}{3d^3} + \frac{c^2}{1-d} - \frac{c^6}{3(1-d)^3}.$$

For $c = \frac{\varepsilon}{2} > 0$, $d_0(y, y_0) < \varepsilon$. Since $d > 1/2$, we can find a bound for two of the terms in the expression for $J(y)$. Thus

$$\frac{c^2}{d} - \frac{c^6}{3d^3} < 2\,\varepsilon^2(1 + \frac{1}{48}\,\varepsilon^4).$$

If we choose d appropriately near 1, then the expression

$$\frac{c^2}{1-d} - \frac{c^6}{3(1-d)^3}$$

can be made negative and, in absolute value, arbitrarily large. There are, therefore, admissible functions y having arbitrarily small distances d_0 from y_0 but whose variational integral $J(y)$ is smaller than $J(y_0) = 0$. It follows that y_0 cannot be a strong solution of the variational problem.

6

The Hilbert independence integral and sufficiency conditions

An admissible function y_0 is a solution of the variational problem if, for all admissible functions y, it holds that

$$J(y) \geq J(y_0); \quad \text{i.e.} \quad J(y) - J(y_0) \geq 0. \tag{6.1}$$

A direct estimate is possible, naturally, in only a few cases. In this section we will derive a condition which arises out of the estimate (6.1). This sufficient condition makes use of a path-independent integral.

(6.2) Definition
An integral I^* is said to be *path-independent* if, for two arbitrary functions y_1 and y_2, whose graphs have the same initial point $(x_a, y_1(x_a)) = (x_a, y_2(x_a))$ and the same endpoint $(x_b, y_1(x_b)) = (x_b, y_2(x_b))$, $I^*(y_1) = I^*(y_2)$. The value of the integrand I^* depends, therefore, only on the integral and the endpoints of the curve y, not on its path.

A trivial example for a path- independent integral is the integral with the integrand $f(x, y, y') = y'$,

$$I^*(y) = \int_{x_a}^{x_b} y'(x)\, dx = y(x_b) - y(x_a) \quad \text{for} \quad y : [x_a, x_b] \mapsto \mathbb{R}.$$

The basic idea of the sufficiency condition here is the following: Suppose I^* (with integrand f^*) is a path-independent integral which provides the same value as the variational integral J; that is,

$$I^*(y_0) = J(y_0) \tag{6.3}$$

for the admissible functions being investigated. Since I^* is path-independent and since all admissible functions satisfy the boundary conditions, it follows that

$$I^*(y_0) = I^*(y)$$

for all admissible functions y. Thus the difference $J(y) - J(y_0)$ can be written

$$\begin{aligned} J(y) - J(y_0) &= J(y) - I^*(y_0) = J(y) - I^*(y) \\ &= \int_a^b [f(x, y(x), y'(x)) - f^*(x, y(x), y'(x))] \, dx. \end{aligned} \tag{6.4}$$

Obviously, $J(y) - J(y_0) \geq 0$ if for all $x \in [a, b]$ the integrand of the above integral is not less than zero. Thus, if for the given admissible function y_0 we can find a path-independent integral I^* with integrand f^* which satisfies the conditions

$$f(x, y_0(x), y_0'(x)) = f^*(x, y_0(x), y_0'(x)) \quad \text{for all} \quad x \in [a, b] \tag{6.5}$$

and

$$\begin{aligned} &f(x, y, y') \geq f^*(x, y, y') \ \text{ for all } \ (x, y, y') \\ &\text{in the domain of definition of } \ f \ \text{ and } \ f^*, \end{aligned} \tag{6.6}$$

then y_0 will be a solution of the variational problem. In summary, (6.3) follows from (6.5) and, because of (6.6), the value of the integral (6.4) is non-negative. In what follows we will derive a path-independent integral with property (6.5).

6.1 FIELDS OF EXTREMALS

In order to construct an appropriate path-independent integral I^*, a certain one-parameter family of extremals is needed. We will call this family a *field of extremals.*

(6.7) Definition
Let M be an open set in the $x\alpha$-plane. Then the functions $\tilde{y} : M \mapsto \mathbb{R}$, $(x, \alpha) \mapsto \tilde{y}(x, \alpha)$ form a *family of extremals* if

(i) for each α, the function $x \mapsto \tilde{y}(x, \alpha)$ is an extremal; and
(ii) each function $\tilde{y}$ has continuous partial derivatives with respect to x up to the fourth derivative, has at least a continuous first partial with respect to α, and has continuous mixed second partials.

Furthermore, such a family of extremals is called a *field of extremals*, and a region G,

$$G = \{(x, \tilde{y}(x, \alpha)) : (x, \alpha) \in M\},$$

of the xy-plane which is covered by the family of extremals $\tilde{y}$, is called a *field* if:

(iii) for all $(x, \alpha) \in M$, $\tilde{y}_\alpha(x, \alpha) \neq 0$; and if
(iv) to each point $(x_0, y_0) \in G$ there corresponds one and only one extremal $\tilde{y}(\cdot, \alpha_0)$ of the family $\tilde{y}$ which passes through (x_0, y_0) and for which $\tilde{y}(x_0, \alpha_0) = y_0$ is also valid.

A field of extremals $\tilde{y}$ is called a *central field* if all extremals of the family $\tilde{y}$ pass through a single point which, of course, does not belong to G.

Property (iv) can also be formulated as follows: For each $(x, y) \in G$ the equation $y = \tilde{y}(x, \alpha)$ can be solved uniquely for α to obtain a function $\hat{\alpha}$, which, because of (iii), has continuous partial derivatives with respect to x and y. Thus we may write

$$y = \tilde{y}(x, \hat{\alpha}(x, y)).$$

The slope of the field of extremals $\tilde{y}$ at the point (x_0, y_0) is defined to be the slope $p(x_0, y_0)$ of the extremal which passes through (x_0, y_0):

$$p(x_0, y_0) = \tilde{y}_x(x_0, \hat{\alpha}(x_0, y_0)).$$

An extremal y_0 is said to be *embedded in the extremal field* $\tilde{y}$ if there is an α-value α_0 such that for all x

$$y_0(x) = \tilde{y}(x, \alpha_0).$$

We illustrate these concepts and definitions with examples.

Example 1. The integrand of the variational problem may depend only on y'. If so, then the extremals are of the form

$$y(x) = \alpha_1 x + \alpha_2,$$

where α_1 and α_2 are the integration constants. The family

$$\tilde{y}(x, \alpha) = \alpha\, x, \qquad (x, \alpha) \in M = \{(x, \alpha) \in \mathbb{R}^2 : x > 0\}$$

is a field of extremals, indeed, a central field, which covers the region

$$G = \{(x, y) \in \mathbb{R}^2 : x > 0\}.$$

For each $(x_0, y_0) \in G$ there is exactly one $\alpha_0 = y_0/x_0$ for which the extremal $\tilde{y}(\cdot, \alpha_0)$ passes through (x_0, y_0). The slope of the field of extremals $\tilde{y}$ at (x_0, y_0) is given by

$$p(x_0, y_0) = \hat{\alpha}(x_0, y_0) = \frac{y_0}{x_0}.$$

It can be easily checked that $\tilde{y} : \mathbb{R}^2 \mapsto \mathbb{R}, \tilde{y}(x, \alpha) = x + \alpha$ is also a field of extremals for the variational problem. It is not, however, a central field, and in this case, $G = \mathbb{R}^2$.

Example 2. Consider the problem of the shortest curve joining two points on the sphere (cf. Section 2.3). The integrand f of the variational problem is

$$f(x, y, y') = \sqrt{\cos^2 y + y'^2},$$

and the extremals have been shown to be

$$y(x) = \arctan\left[\frac{\sqrt{1-c^2}}{c}\sin(x+x_0)\right],$$

where c and x_0 are constants of integration. The family

$$\tilde{y}(x,\alpha) = \arctan\left[\frac{\alpha}{\sqrt{1-\alpha^2}}\sin x\right],$$

where $(x,\alpha) \in M = (0,\pi)\times(-1,1)$, is a field of extremals which forms a central field. Clearly, $\tilde{y}_\alpha(x,\alpha) \neq 0$ for $x \in (0,\pi)$. Also, through each point $(x_0,y_0) \in G = (0,\pi)\times(-\pi/2,\pi/2)$ there passes precisely one extremal of the family $\tilde{y}$, namely the extremal $\tilde{y}(\cdot,\alpha_0)$ where

$$\alpha_0 = \frac{\tan y_0}{\sqrt{\sin^2 x_0 + \tan^2 y_0}}.$$

The extremal y_0 with $y_0(x) = 0$ is embedded in this field of extremals. From the above derivation it is evident that the domain of definition of M could not have been chosen larger. All of the extremals of the family of extremals pass through the points $(0,0)$ and $(\pi,0)$.

6.2 THE HILBERT INTEGRAL AND THE WEIERSTRASS SUFFICIENT CONDITION

With these preliminaries we can define the path-independent *Hilbert integral.* It will lead us to a sufficient condition for solutions of variational problems in which the Weierstrass excess function appears. We will then apply this condition to an example.

(6.8) Definition
Let $\tilde{y}$ be a field of extremals; let $p(x,y)$ be the slope of $\tilde{y}$ at the point (x,y) and let G be the field belonging to $\tilde{y}$. The *Hilbert integral* is defined to be

$$I^* = \int A(x,y)\,dx + B(x,y)\,dy,$$

where

$$A(x,y) = f(x,y,p(x,y)) - p(x,y) f_{y'}(x,y,p(x,y)),$$
$$B(x,y) = f_{y'}(x,y,p(x,y)).$$

Thus, for a piecewise smooth function $y : [x_a, x_b] \mapsto \mathbb{R}$, whose graph lies in G,

$$I^*(y) = \int_{x_a}^{x_b} [A(x,y(x)) + B(x,y(x))\,y'(x)]\,dx.$$

The integrand f^* of I^* is defined to be

$$f^*(x,y,y') = A(x,y) + B(x,y)\,y'.$$

(6.9) Definition
A subset G of $\mathbb{R}^2$ is said to be *connected* if any two points in G can be joined by a continuous curve which lies entirely within G. An open and connected set $G \subset \mathbb{R}^2$ is said to be *simply connected* if the interior of every closed curve having no multiple points and lying entirely in G is also in G.

(6.10) Theorem
Let $\tilde{y}$ be a field of extremals whose field G is open, connected and simply connected. Then the Hilbert integral with respect to $\tilde{y}$ is path-independent.

We defer the proof of this theorem until Section 6.4. We want to see first of all what use the introduction of fields of extremals and of the path-independent integral has for the calculus of variations. The particulars of the proof of (6.10) are not required for an understanding of the following Weierstrass sufficiency condition. Those, however, who wish to adhere to the sequence 'theorem – proof – consequences of the theorem' or who would like to know why the field G in (6.10) is assumed to be simply connected, should read Section 6.4 next, and then proceed to the following paragraphs.

Let $y_0 : [a, b] \mapsto \mathbb{R}$ be an extremal and an admissible function of the stated variational problem. We will set down conditions under which y_0 is a solution of the problem. We assume that y_0 is embedded in a field of extremals $\tilde{y}$. We form the Hilbert integral I^* for this field of extremals. First we make sure that the requirement of (6.5) – that is, that $f^*(x, y_0(x), y_0'(x)) = f(x, y_0(x), y_0'(x))$ – is satisfied so that $I^*(y_0) = J(y_0)$

$$\begin{aligned} f^*(x, y_0(x), y_0'(x)) &= A(x, y_0(x)) + B(x, y_0(x))\, y_0'(x) \\ &= f(x, y_0(x), p(x, y_0(x))) - p(x, y_0(x))\, B(x, y_0(x)) \\ &\quad + y_0'(x)\, B(x, y_0(x)). \end{aligned}$$

But the slope $p(x, y_0(x))$ of the field of extremals $\tilde{y}$ at the point $(x, y_0(x))$ is simply $y_0'(x)$, since y_0 is the (uniquely determined) extremal of the family which passes through $(x, y_0(x))$. Thus, (6.5) is satisfied.

Now let y be an arbitrary admissible function of the variational problem whose graph $\{(x, y(x)) : x \in [a, b]\}$ lies in the field G. If G is open, connected and simply connected, then according to (6.10) I^* is path-independent and

$$J(y) - J(y_0) = J(y) - I^*(y_0) = J(y) - I^*(y),$$

since y and y_0 have the same boundary conditions. Further,

$$\begin{aligned} J(y) - I^*(y) &= \int_a^b [f(x, y(x), y'(x)) - f^*(x, y(x), y'(x))]\, dx \\ &= \int_a^b [f(x, y(x), y'(x)) - f(x, y(x), p(x, y(x))) \end{aligned}$$

$$- f_{y'}(x, y(x), p(x, y(x)))(y'(x) - p(x, y(x)))]\, dx$$
$$= \int_a^b \mathcal{E}(x, y(x), p(x, y(x)), y'(x))\, dx,$$

where $\mathcal{E}$ is the Weierstrass excess function defined in (4.4). This brings us to our goal: $J(y) - J(y_0) = J(y) - I^*(y) \geq 0$, provided the integrand of the integral above does not take on negative values for $x \in [a, b]$. We have proved the Weierstrass sufficiency condition.

(6.11) Theorem
Let y_0 be an admissible function of the given variational problem $J(y) \to \min$ and let there be given a field of extremals $\tilde{y}$ in which y_0 is embedded. Let the field G of $\tilde{y}$ be open, connected and simply connected, and suppose that all admissible functions y lie in G; that is, their graphs lie in G. Furthermore, suppose that the function

$$(x, y, p) \in G \times \mathbb{R} \mapsto \mathcal{E}(x, y, p(x, y), q) \in \mathbb{R},$$

where p is the slope of the field, has no negative values. Then y_0 is a solution of the variational problem.

This theorem establishes the fact that the line segment joining two points A and B in the plane is the shortest curve joining the points. (Exercise: Show this using the field of extremals given in the first example of Section 6.1.)

Theorem (6.11) is also decisive for the example of the shortest curve joining two points A and B on the sphere. (The formulation of this problem and the calculation of the extremals are found in Section 2.3; a field of extremals $\tilde{y}$ which we can use is given in the second example of the last section (6.1).) Let A and B be given by means of their respective longitude and latitude: $A = (a, 0), B = (b, 0)$. The equator $y_0 = 0$ which joins A and B is embedded in the field of extremals; the field G is $G = (0, \pi) \times (-\pi/2, \pi/2)$. G also satisfies the assumptions of Theorem (6.11). Each admissible function of the problem also lies in G if $0 < a < b < \pi$. Since the Weierstrass excess function

$$\mathcal{E}(x,y,p,q) = (\cos^2 y + p^2)^{-1/2}[(\cos^2 y + q^2)^{1/2}(\cos^2 y + p^2)^{1/2} - pq - \cos^2 y] \geq 0$$

for all (x,y,p,q), all the conditions of Theorem (6.11) are fulfilled and, consequently, the equator $y_0 = 0$ is the solution of the variational problem for $0 < a < b < \pi$. Recall that we had already recognized in Section 5.2 (second example) that for points $A = (0,0)$ and $B = (b,0)$, $b > \pi$, the equator y_0 is not the shortest curve joining A and B. It should be noted that in order to solve a variational problem it is not essential that one first check the necessary conditions as summarized in Section 5.4. If one has found an extremal y_0 for the variational problem which satisfies the boundary conditions, and if one can show that the extremal satisfies the conditions of Theorem (6.11), then a solution to the variational problem has already been found.

6.3 FURTHER SUFFICIENT CONDITIONS

A critical glance at the requirements of the Weierstrass sufficiency condition (6.1) shows that in many cases there are problems:

1. The family of extremals in which y_0 shall be embedded must be determined.
2. Property (iv) for the field of extremals must be shown to hold, and the field G must be determined; to do this, one must solve the equation $y = \tilde{y}(x,\alpha)$ for α.

We are interested, therefore, in conditions which can be more easily checked and from which the existence of an appropriate field follows. The conditions which we provide, however, will guarantee only local fields of extremals – that is, fields of extremals which cover only a certain neighbourhood of the graph of y_0. Since the estimate of $J(y) - J(y_0)$, however, is valid only for admissible functions y whose graphs lie in the field, the sufficient conditions which are obtained from local fields of extremals are only conditions for strong or weak solutions.

(6.12) Theorem
Let y_0 be a regular extremal for the aforementioned variational problem and let y_0 satisfy the given boundary conditions. Suppose also that in $(a, b]$ there is for y_0 no point conjugate to $x = a$ and that

$$\mathcal{E}(x, y, y', q) \geq 0$$

for all $q \in \mathbb{R}$ and for all triples (x, y, y') which are sufficiently near the set $\{(x, y_0(x), y_0'(x)) : x \in [a, b]\}$. In other words, there is an $\varepsilon > 0$ such that $\mathcal{E}(x, y, y', q) \geq 0$ for all $q \in \mathbb{R}$ and for all triples (x, y, y') for which $x \in [a, b]$, $|y - y_0(x)| < \varepsilon$ and $|y' - y_0'(x)| < \varepsilon$. Then y_0 is a strong solution of the variational problem.

Proof
Since we want to use the method of proof employed in Theorem (6.11), we must show that there is an extremal field in which y_0 is embedded and which covers a certain ε^*-band (cf. Fig. 1.6 in Chapter 1) about the graph of y_0. We will show that the extremals through a fixed point (centre) left of $x = a$ form such a field of extremals. This field does not need to be calculated here. It suffices to be convinced of its existence. Since y_0 is regular, $f_{y'y'}(x, y_0(x), y_0'(x)) \neq 0$ for all $x \in [a, b]$. The extremal y_0 may be extended as an extremal of the variational problem to the interval $[a - \delta_1, b]$, $\delta_1 > 0$. We designate this extended extremal with y_0 also. Because $f_{y'y'}$ is continuous, there is a $\delta > 0$ such that $f_{y'y'}(x, y, y') \neq 0$ for all triples (x, y, y') with $a - \delta \leq x \leq b$, $|y - y_0(x)| < \delta$ and $|y' - y_0'(x)| < \delta$. In this xyy'-region the Euler equation is solvable for y'', and the initial value problem

consisting of the Euler equation and the initial conditions

$$\begin{gathered} y(a-\varepsilon)=y_0(a-\varepsilon), \quad y'(a-\varepsilon)=y_0'(a-\varepsilon)+\alpha, \\ \text{where } \varepsilon \text{ is a parameter and } 0 \le \varepsilon < \delta, \end{gathered} \tag{6.13}$$

is solvable and yields for each fixed ε a family of extremals

$$\tilde{y} : (x, \alpha) \mapsto \tilde{y}(x, \alpha).$$

This family contains the extremal y_0 $(\alpha = 0)$; $\tilde{y}(x, \alpha)$ depends, of course, on ε. The parameter ε will be determined appropriately later.

For each fixed ε the maximal domain of definition for the family of extremals is a set M_ε in the $x\alpha$-plane which contains a neighbourhood of the $x\alpha$-values $\{(x, 0) : x \in [a - \varepsilon, b]\}$ belonging to y_0. Now from the theorems about the independence of the solutions of an initial value problem from the parameters (here α and ε), it follows that each function $\tilde{y}$ satisfies the differentiability conditions required in Definition (6.7)(ii). Condition (iii) of (6.7) requires that $\tilde{y}_\alpha(x, \alpha) \neq 0$ hold for all (x, α) in the domain of definition of the field of extremals. According to Theorem (5.11), a solution of the Jacobi equation having a continuous second derivative is given by $Y(x) = \tilde{y}_\alpha(x, 0)$. From the initial conditions (6.13) it follows that

$$Y(a-\varepsilon) = 0, \;\; Y'(a-\varepsilon) = \tilde{y}_{\alpha x}(a-\varepsilon, 0) = \tilde{y}_{x\alpha}(a-\varepsilon, 0) = 1.$$

Thus Y is not the null function. Since $f^0_{y'y'} \neq 0$ for all $x \in [a - \delta, b]$, the Jacobi equation is regular so that Y does not vanish on any interval of positive length. Since we assumed that $x = a$ has no conjugate point in $(a, b]$, it follows that the function Y, which is obtained when the parameter ε has the value zero, has no zero in $(a, b]$. Also, since $Y(x)$ and $\tilde{y}(x, \alpha)$ are continuous with respect to the parameters ε and α, respectively, there is an $\varepsilon = \varepsilon_0$, $0 < \varepsilon_0 < \delta$, and an $\alpha_0 > 0$ such that the family of extremals $\tilde{y}$ determined by setting $\varepsilon = \varepsilon_0$ has the following properties:

$$Y(x) = \tilde{y}_\alpha(x, 0) \neq 0 \quad \text{for all} \quad x \in (a - \varepsilon_0, b]$$

and

$$\tilde{y}_\alpha(x, \alpha) \neq 0 \text{ for all } x \in (a - \varepsilon_0, b] \text{ and all } \alpha \text{ with } |\alpha| < \alpha_0.$$

Now for $\tilde{y}$ property (iv) follows from property (iii). For according to the implicit function theorem, the equation $y = \tilde{y}(x, \alpha)$ can be solved uniquely for α. Then the assumption that there is a point (x_0, y_0) through which two extremals pass, i.e.

$$y_0 = \tilde{y}(x_0, \alpha_1) = \tilde{y}(x_0, \alpha_2),$$

where $\alpha_1 < \alpha_2$, $|\alpha_1| < \alpha_0$ and $|\alpha_2| < \alpha_0$, leads to a contradiction since

$$\tilde{y}(x_0, \alpha_2) - \tilde{y}(x_0, \alpha_1) = \tilde{y}_\alpha(x_0, \alpha^*)(\alpha_2 - \alpha_1) \neq 0,$$

where α^* is a value between α_1 and α_2. This concludes the proof of the existence of a field of extremals $\tilde{y}$ in which y_0 is embedded.

The domain of definition for $\tilde{y}$ can be chosen so that the field G for $\tilde{y}$ is open, connected and simply connected, and so that it contains an ε^*-band about y_0. The slope function $p(x, y)$ for $\tilde{y}$ at the point $(x, y) \in G$ differs little from the slope of the field of extremals at $(x, y_0(x))$, so it is sufficient if for some $\varepsilon > 0$ the Weierstrass $\mathcal{E}$-function is not negative for the values (x, y, y', q) with $x \in [a, b]$, $|y - y_0(x)| < \varepsilon$ and $|y' - y_0'(x)| < \varepsilon$, and if $q \in \mathbb{R}$ is not negative. This concludes the proof of Theorem (6.12).

(6.14) Lemma

If in Theorems (6.11) *and* (6.12) *the assumption* $\mathcal{E} \geq 0$ *is sharpened, respectively, to*

$$\begin{aligned} &\mathcal{E}(x, y, p(x, y), q) > 0 \textit{ for all } (x, y) \in G \\ &\quad \textit{and all } q \in \mathbb{R} \textit{ with } q \neq p(x, y) \end{aligned} \tag{6.15}$$

and

$$\begin{aligned} &\mathcal{E}(x, y, y', q) > 0 \quad \textit{for all} \quad (x, y, y', q) \\ &\textit{with} \quad x \in [a, b], |y - y_0(x)| < \varepsilon, |\, y' - y_0'(x) \,| < \varepsilon \\ &\textit{and} \quad q \in \mathbb{R} \quad \textit{with} \quad q \neq y', \varepsilon > 0, \end{aligned} \tag{6.16}$$

then y_0 is a proper, or proper strong, solution of the variational problem.

Proof

That y_0 is a solution or a strong solution follows from Theorems (6.11) and (6.12), respectively. It needs to be shown that y_0 is the only solution or the uniquely determined local solution. Thus, suppose that y^* is an admissible function for which it holds that $J(y^*) = J(y_0)$, and whose graph lies in the field we constructed for y_0. From

$$J(y^*) - J(y_0) = \int_a^b \mathcal{E}(x, y^*(x), p(x, y^*(x)), y^{*\prime}(x))\, dx = 0,$$

and the fact that $\mathcal{E}(x, y, p(x, y), y') > 0$ for $y' \neq p(x, y)$ it follows that

$$y^{*\prime}(x) = p(x, y^*(x)) \quad \text{for all} \quad x \in [a, b]. \tag{6.17}$$

But (6.17) is the first-order differential equation whose solutions are precisely the extremals of the field of extremals; y^* is therefore an extremal. Since y^* satisfies the boundary conditions $y^*(b) = y_b = y_0(b)$, y^* must agree with y_0 everywhere, for there is only one extremal of the field of extremals that passes through (b, y_b). We conclude that y_0 is the uniquely determined solution or relative solution of the variational problem.

(6.18) Lemma

Suppose the assumptions of Theorem (6.12) are satisfied, except for the requirement $\mathcal{E} \geq 0$, instead of which it is required that there exists an $\varepsilon > 0$ such that

$$f_{y'y'}(x, y, y') > 0 \quad \textit{for all} \quad (x, y, y'), \tag{6.19}$$

where

$$x \in [a, b], \quad |y - y_0(x)| < \varepsilon \quad \textit{and} \quad | y' - y_0'(x) | < \varepsilon.$$

Then y_0 is a weak solution of the variational problem.

Proof

To prove this lemma we construct a field of extremals $\tilde{y}$ as in the proof of Theorem (6.12); thus y_0 shall be embedded in the field of extremals and $\tilde{y}$ shall cover a neighbourhood of the graph of y_0 in the xy-plane. Now let y be an admissible function whose distance d_1 from y_0 is sufficiently small; i.e. $y(x)$ and $y'(x)$ differ little from y_0 and y_0', respectively. Also let y lie in the field. Then it follows that

$$J(y) - J(y_0) = \int_a^b \mathcal{E}(x, y(x), p(x, y(x)), y'(x))\, dx. \qquad (6.20)$$

Suppose now that we regard x and y as fixed and consider $\mathcal{E}$ as a function of the final two variables. If $\mathcal{E}$ is now expanded about the point $y = p(x, y)$, we receive

$$\begin{aligned} &\mathcal{E}(x, y(x), p(x, y(x)), y'(x)) \\ &= \frac{1}{2} f_{y'y'}(x, y(x), p^*)(y'(x) - p(x, y(x)))^2, \end{aligned}$$

where $p^* = t\,p(x, y(x)) + (1-t)y'(x)$, $0 \le t \le 1$, is an intermediate value for p. The integrand of (6.20) is greater than or equal to zero (cf. (6.19)) if the d_1- distance of y from y_0 is sufficiently small.

We use the sufficient condition of Theorem (6.12) to investigate an example.

Example

Recall the variational problem (1.2) of the smallest surface of revolution:

$$J(y) = \int_a^b y(x)\sqrt{1 + y'^2(x)}\, dx \to \min,$$

$$y(x) > 0 \quad \text{for} \quad x \in [a, b], \quad y(a) = y_a, \quad y(b) = y_b.$$

We have already found the extremals of this problem in Section 2.3:

$$y^*(x; c, d) = \frac{1}{c}\cosh(cx + d). \qquad (6.21)$$

Here c and d are integration constants, $c > 0$. We assume also that there is an extremal

$$y_0(x) = \frac{1}{c_0}\cosh(c_0 x + d_0), \quad y_0(a) = y_a, \quad y_0(b) = y_b,$$

which satisfies the stated boundary conditions. (Recall here the discussion of the boundary value problem in Section 2.3.) For all values in the domain of definition, the Weierstrass $\mathcal{E}$-function is non-negative, and it holds that $f_{y'y'}(x, y, y') > 0$ for all x, y, y' (cf. Section 4.3). We must also investigate whether there are conjugate points for $y_0(x)$ with respect to the point $x = a$; that is, whether there are zeros $c > a$ of those solutions $Y \neq 0$ of the Jacobi equation which satisfy the initial condition $Y(a) = 0$. We need not actually determine the Jacobi equation since, according to Theorem (5.14), Y_1 and Y_2, defined by

$$Y_1(x) = y_c^*(x, c_0, d_0) = \frac{-1}{c_0^2}\cosh(c_0 x + d_0) + \frac{x}{c_0}\sinh(c_0 x + d_0)$$

and

$$Y_2(x) = y_d^*(x, c_0, d_0) = \frac{1}{c_0}\sinh(c_0 x + d_0),$$

are basis solutions (vectors) of the Jacobi equation. The general solution of the Jacobi equation may be represented as a linear combination $Y = C_1 Y_1 = C_2 Y_2$ with constants C_1 and C_2. We obtain the conditions on C_1 and C_2 from the initial conditions

$$\begin{aligned} Y(a) = C_1 &\left[\frac{a}{c_0}\sinh x_a - \frac{1}{c_0^2}\cosh x_a\right] \\ &+ C_2 \frac{1}{c_0}\sinh x_a = 0, \end{aligned} \tag{6.22}$$

where $x_a = c_0 a + d_0$. We must distinguish here between two cases:

Case 1: $x_a = 0$; Case 2: $x_a \neq 0$.

In the first case, it follows from (6.22) that $C_1 = 0$, so that only the function $Y = C_2 Y_2$, with $C_2 \in \mathbb{R}$, can satisfy the initial conditions $Y(a) = 0$. But Y_2 has no zeros for $x > a$. Thus, there are no points conjugate to a. According to Theorem (6.12), y_0 is a strong solution of the variational problem. The condition $x_a = 0$ means that a is the abcissa of the lowest point on the catenary k_0 defined by

$$x \in \mathbb{R} \mapsto (x, \frac{1}{c_0} \cosh(c_0 x + d_0)).$$

In the second case, since $\sinh x_a \neq 0$ and because of (6.22),

$$C_2 = -C_1(a - \frac{1}{c_0} \coth x_a).$$

A function $Y \neq 0$ with the initial condition $Y(a) = 0$ has a zero at $x = c > a$; that is, precisely where

$$c_0 c \sinh x_c - \cosh x_c - (ac_0 - \coth x_a) \sinh x_c = 0,$$

with $c_0 c + d_0$ abbreviated here by x_c. This condition is not satisfied if $\sinh x_c = 0$. So we may divide with $\sinh x_c$ and obtain

$$\coth x_c - c_0 c = \coth x_a - c_0 a.$$

To solve this equation we make use of the graph of the function F, $F(x) = \coth(c_0 x + d) - c_0 x$ (see Fig. 6.1). Clearly, $x_0 = -d_0/c_0$ is a pole of F. Otherwise, in the intervals $(-\infty, x_0)$ and (x_0, ∞), F is monotone and one-to-one. In the event that $a < x_0$, there is precisely one value $c > x_0$ such that $F(a) = F(c)$. This point c is conjugate to a. If, however, $a > x_0$, then there is no value $c > a$ for which $F(c) = F(a)$ and hence no point conjugate to $x = a$. The condition $a < x_0$ means that the left boundary point of the graph of y_0 lies left of the lowest point on the catenary k_0, a part of which is represented by y_0. In the case where $a > x_0$, y_0 is a strong solution for arbitrary b. If $a < x_0$, then y_0 is a strong solution provided $b < c$. If $b > c$, however, then y_0 does not satisfy the necessary Jacobi condition and y_0 cannot even be a weak solution of the variational problem. In

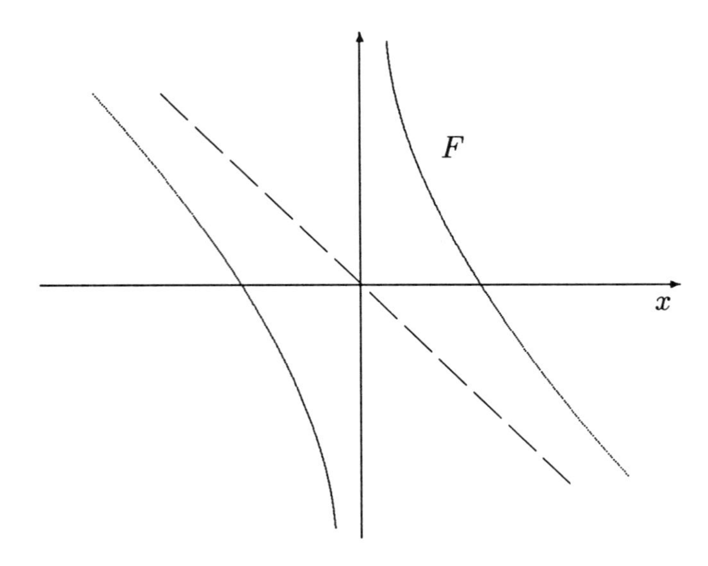

Fig. 6.1. A graphical solution.

this case there is a second extremal which satisfies the boundary conditions and provides a strong solution.

It can be shown additionally that for the problem of the smallest surface of revolution, $\mathcal{E}(x, y, y', q)$ is greater than zero if $q \neq y'$. Thus, according to Lemma (6.14), y_0 is a proper strong solution provided a and b satisfy one of the following conditions:

$$\begin{aligned} &\text{(i)} \quad a \geq -d_0/c_0 \quad \text{or} \\ &\text{(ii)} \quad a < -d_0/c_0 \quad \text{and} \quad b < c, \end{aligned}$$

where c_0 and d_0 are the integration constants for the extremal y_0 and c denotes the point conjugate to a.

Exercises

(6.23) Among all piecewise smooth functions y which satisfy the boundary conditions

$$y(0) = -1, \quad y(\pi/4) = 0, \quad b = \pi/4,$$

find those which minimize the integral

$$\int_0^b (y'^2(x) - 4y^2(x) - 8y(x))\, dx.$$

(6.24) Find the solution of the variational problem

$$\int_a^b \frac{y(x)}{y'^2(x)}\,dx \to \min, \qquad y(a) = y_a, \qquad y(b) = y_b,$$

where y_a and y_b are positive constants.

6.4 PROOF OF THE PATH-INDEPENDENCE OF THE HILBERT INTEGRAL

The existence of a path-independent integral I^* was of fundamental significance for the comparison of $J(y)$ and $J(y_0)$. In this section we concern ourselves with the proof of Theorem (6.10) which says that the Hilbert integral I^*, formed with the help of the field of extremals $\tilde{y}$, is path-independent, provided the field G belonging to $\tilde{y}$ is open, connected and simply connected. Recall that

$$I^* = \int A(x,y)\,dx + B(x,y)\,dy,$$

where

$B(x,y) = f_{y'}(x,y,p(x,y))$,
$A(x,y) = f(x,y,p(x,y)) - p(x,y)\,B(x,y)$, and
$p(x,y)$ is the slope function of the field of extremals $\tilde{y}$ at the point (x,y).

Thus, for a piecewise smooth function $y : [x_1, x_2] \mapsto \mathbb{R}$, whose graph lies in the field G of $\tilde{y}$,

$$I^*(y) = \int_{x_1}^{x_2} [A(x,y(x)) + B(x,y(x))\,y'(x)]\,dx. \tag{6.25}$$

The integral I^* is path-independent if for all y the value of the integral depends only on the boundary values of y, not on the path of the graph between the endpoints. If there exists a function $P : G \mapsto \mathbb{R}$ ('Potential') having continuous first and second derivatives so that

$$I^* = \int dP = \int P_x(x,y)\,dx + P_y(x,y)\,dy$$

for the piecewise smooth function y, that is, if

$$\begin{aligned}
I^*(y) &= \int_{x_1}^{x_2} [P_x(x, y(x)) + P_y(x, y(x))\, y'(x)]\, dx \\
&= \int_{x_1}^{x_2} \frac{d}{dx}[P(\cdot, y)](x)\, dx \qquad (6.26)\\
&= P(x_2, y(x_2)) - P(x_1, y(x_1)),
\end{aligned}$$

then I^* is clearly path-independent. The integrals (6.25) and (6.26) have the same structure. Thus we seek a function P, such that for all $(x, y) \in G$:

$$\begin{aligned}
\text{(i)}\quad P_x(x, y) &= A(x, y) \\
&= f(x, y, p(x, y)) - p(x, y)\, f_{y'}(x, y, p(x, y)); \\
\text{(ii)}\quad P_y(x, y) &= B(x, y) = f_{y'}(x, y, p(x, y)). \qquad (6.27)
\end{aligned}$$

We may regard (6.27) as a system of partial differential equations of first order for the function P which we seek. According to the theorem about the solubility of partial differential equations there is, for each prescribed initial value $P(x_a, y_a) = P_a$, a uniquely determined local solution P of the system (6.27) which has continuous second partials, provided the following conditions are satisfied:

(i) The functions A and B on the right side of (6.27) are continuous functions of x and y which also have continuous partial derivatives; and

(ii) for all $(x, y) \in G$,

$$A_y(x, y) = B_x(x, y) \quad \text{(integrability condition)}. \qquad (6.28)$$

Condition (i) is clearly satisfied. We must proceed carefully, however, in order to verify (ii) because the functions P, which belong to the field of extremals, as well as $\hat{\alpha}$ (cf. Definition (6.7)) and the relationship between $\tilde{y}$, p and $\hat{\alpha}$, must be regarded.

Consider next the equations

$$A_y(x,y) = f_y - p\,f_{y'y} - p\,f_{y'y'}\,p_{y'},$$
$$B_x(x,y) = f_{y'x} + f_{y'y'}\,p_x.$$

For the partial differentiation here the arguments of f are x, y and $p(x,y)$ while the arguments for p, p_x and p_y are x and y. Since $p(x,y)$ is the slope function for the field $\tilde{y}$, it follows that

$$p(x,y) = \tilde{y}_x(x,\hat{\alpha}(x,y)),$$
$$p_x(x,y) = \tilde{y}_{xx}(x,\hat{\alpha}(x,y)) + \tilde{y}_{x\alpha}(x,\hat{\alpha}(x,y))\hat{\alpha}_x(x,y),$$
$$p_y(x,y) = \tilde{y}_{x\alpha}(x,\hat{\alpha}(x,y))\hat{\alpha}_y(x,y).$$

If we substitute these expressions into $A_y - B_x$, we obtain

$$\begin{aligned} A_y(x,y) - B_x(x,y) &= f_y - f_{y'x} - f_{y'y}\tilde{y}_x \\ &\quad - f_{y'y'}(\tilde{y}_{xx} + \tilde{y}_{x\alpha}\hat{\alpha}_x(x,y) + \tilde{y}_{x\alpha}\hat{\alpha}_y(x,y)\tilde{y}_x). \end{aligned}$$

Arguments for the partial differentiation of f are x, y and $\tilde{y}_x(x,\hat{\alpha}(x,y))$. The arguments for $\tilde{y}$ are x and $\hat{\alpha}(x,y)$.

We may now make use of the fact that $\tilde{y}$ is a field of extremals. Let us check the integrability condition (6.28) at an arbitrary point (x_p,y_p) of the field G. Thus, let y_0 be the uniquely determined extremal of the field of extremals $\tilde{y}$ which passes through (x_p,y_p) and let α_0 be the α-value belonging to y_0. Then

$$y_0(x_p) = y_p, \quad y_0(x) = \tilde{y}(x,\alpha_0), \quad \alpha_0 = \hat{\alpha}(x,y_0(x)),$$
$$y_0'(x) = \tilde{y}_x(x,\alpha_0), \quad y_0''(x) = \tilde{y}_{xx}(x,\alpha_0).$$

If we differentiate $\alpha_0 = \hat{\alpha}(x,y_0(x))$ with respect to x we obtain

$$0 = \hat{\alpha}_x(x,y_0(x)) + \hat{\alpha}_y(x,y_0(x))\,y_0'(x).$$

From this we conclude that

$$\begin{aligned} A_y(x_p,y_p) - B_x(x_p,y_p) &= f_y - f_{y'x} - f_{y'y}y_0'(x_p) - f_{y'y'}y_0''(x_p) \\ &\quad + f_{y'y'}\tilde{y}_{x\alpha}(x,\alpha_0)[\hat{\alpha}_x(x_p,y_p) + \hat{\alpha}_y(x_p,y_p)\,y_0'(x_p)] \\ &= f_y(x_p,y_0(x_p),y_0'(x_p)) - \frac{d}{dx}[f_{y'}(\cdot,y_0,y_0')](x_p) = 0 \end{aligned}$$

since y_0, an extremal, satisfies the Euler equation. This concludes the proof of Theorem (6.28).

The theorem about partial differential equations which was cited above speaks only of a local solution P, that is, of a solution defined only in a neighbourhood of the initial point (x_a, y_a). But for the path-independent integral we need a solution of (6.27) which is defined on all of G. But because (6.27) has a simpler structure than the general system of partial differential equations which the theorem treats, it is possible to obtain other, broader results. It can be shown that the solution P is defined on all of G if G is simply connected.

We will determine $P(x_0, y_0)$ explicitly for an arbitrary point (x_0, y_0) in G. To do this we choose an arbitrary, smooth curve k from (x_a, y_a) to (x_0, y_0) and lying entirely in G. Such a curve can always be found since G is assumed to be open and connected. We represent k parametrically

$$t \in [t_a, t_0] \mapsto (x(t), y(t)) \in \mathbb{R}^2,$$

$$(x(t_a), y(t_a)) = (x_a, y_a), \qquad (x(t_0), y(t_0)) = (x_0, y_0).$$

A solution P of (6.27) satisfies the equation

$$\begin{aligned} \frac{d}{dt} P(x,y)(t) &= P_x(x(t), y(t)) \frac{d}{dt} x(t) + P_y(x(t), y(t)) \frac{d}{dt} y(t) \\ &= A(x(t), y(t)) \frac{d}{dt} x(t) + B(x(t), y(t)) \frac{d}{dt} y(t). \end{aligned}$$

The right-hand side is formed from the known functions A, B, x, y, dx/dt and dy/dt. Thus, these functions can be integrated to obtain

$$\begin{aligned} P(x_0, y_0) &= P(x_a, y_a) + \int_{t_a}^{t_0} \frac{d}{dt} P(x,y)(t)\, dt \\ &= P(x_a, y_a) + \int_{t_a}^{t_0} [A(x(t), y(t))\, \frac{d}{dt} x(t) \\ &\quad + B(x(t), y(t))\, \frac{d}{dt} y(t)]\, dt. \end{aligned} \tag{6.29}$$

Though we will not go into the matter here, it should be noted that the proof that the independence of $P(x_0, y_0)$ from the choice of the arc (curve) connecting (x_a, y_a) and (x_0, y_0) depends essentially on the fact that G is simply connected. Without this assumption the theorem would be false, as examples would show.

7

Variational problems with variable boundaries

Until now we have investigated only variational problems for which the admissible functions y must satisfy *boundary conditions* $y(a) = y_a$ and $y(b) = y_b$. In what follows we will refer to these problems as *fixed boundary problems*. We now want to concern ourselves with variational problems for which at least one of the boundary points of the admissible function is "movable" along a boundary curve. Here is an example: Find the shortest arc (airline route) on the surface of the earth connecting a point A (say Frankfurt) and a given curve (Atlantic coast of North America).

7.1 PROBLEMS HAVING A FREE BOUNDARY POINT

The simplest type of problem here is one for which one of the boundary values, say y_a, is given while the other y_b is arbitrary. That is to say, the right endpoint is free to move along a line parallel to the y-axis. Here all admissible functions have the same domain of definition $[a, b]$. More precisely, we set down the following conditions.

(7.1) Admissibility conditions
Admissible functions shall consist of all piecewise smooth functions $y : [a,b] \mapsto \mathbb{R}$ for which $y(a) = y_a$, y_a fixed, while $y(b)$ is arbitrary, provided that $f(x, y(x), y'(x))$, $f(x, y(x), y_l(x))$ and $f(x, y(x), y_r(x))$ are each defined for all $x \in [a,b]$.

The variational problem with *free right endpoint* is stated thus: Among all functions y which satisfy the admissibility conditions (7.1), find those for which the variational integral

$$J(y) = \int_a^b f(x, y(x), y'(x))\, dx$$

assumes its smallest value.

This problem differs from the fixed boundary problem in that the functional J must now be considered with respect to the function space $\mathcal{F}_1$ defined by (7.1), rather than with respect to the smaller function space $\mathcal{F}_0$ defined by the relaxed admissibility conditions (2.22). As before, functions which provide minima for the functionals defined on $\mathcal{F}_1$ will be called *solutions* of the problem. It is also meaningful to speak of relative minima for J on $\mathcal{F}_1$ since the distances d_0 and d_1 (cf. (1.11) and (1.12)) are also defined for functions in $\mathcal{F}_1$. Relative minima of J on $\mathcal{F}_1$ with respect to the distance d_0 are called strong solutions, while relative minima of J with respect to the distance d_1 are called weak solutions of the variational problem.

Now let $y_0 \in \mathcal{F}_1$ be a weak or strong solution, respectively, of the problem with free right endpoint. We call the fixed endpoint problem with the functional J, and for which the boundary conditions $y(a) = y_0(a) = y_a$ and $y(b) = y_0(b) = y_b$ are given, the *fixed endpoint problem belonging to* y_0. Since each admissible function for the fixed endpoint problem belonging to y_0 also satisfies the admissibility conditions (7.1), the following theorem can be stated.

(7.2) Theorem
A weak or strong solution y_0 of the variational problem with free right endpoint satisfies, respectively, all the necessary conditions for a weak or strong solution of the fixed boundary problem belonging to y_0.

Thus, y_0 satisfies the Erdmann corner conditions at the corners, as well as the Euler equation and the necessary conditions of Legendre, Weierstrass and Jacobi. But there is still another necessary condition which must be added and which we will now derive. To this end, we consider a family of admissible functions y,

$$y(x,\alpha) = y_0(x) + \alpha Y(x), \qquad |\alpha| < a_0,$$

where Y is an arbitrary piecewise smooth function with boundary conditions $Y(a) = 0$ and $Y(b)$ arbitrary. We define the function $\Phi : [-\alpha_0, \alpha_0] \mapsto \mathbb{R}$,

$$\Phi(\alpha) = \int_a^b f(x, y(x,\alpha), y_x(x,\alpha))\, dx,$$

which has a relative minimum at the point $\alpha = 0$ since $y_0 = y(\cdot, 0)$ is a weak solution of the problem. Thus $\Phi'(0) = 0$,

$$\Phi'(0) = \int_a^b [f_y^0(x)Y(x) + f_{y'}^0(x)Y'(x)]\, dx$$

(cf. (2.4) for the definition of f_y^0 and $f_{y'}^0$). By partial integration of the second integrand we obtain

$$\begin{aligned}\Phi'(0) = &\int_a^b \left[f_y^0(x) - \frac{d}{dx} f_{y'}^0(x)\right] Y(x)\, dx \\ &+ f_{y'}(b, y_0(b), y_0'(b))\, Y(b) \\ &+ \sum_i [f_{y'}(a_i, y_0(a_i), y_{0l}'(a_i)) \\ &- f_{y'}(a_i, y_0(a_i), y_{0r}'(a_i))] Y(a_i),\end{aligned}$$

where $a_i, i = 1,\ 2,\ \ldots,\ m$, are corners of the function y, and so also of y_0 or Y. Since according to Theorem (7.2) y_0 satisfies the

Euler equation and the corner conditions, the integral and the terms of the sum on the right side vanish so that there remains only

$$\Phi'(0) = 0 = f_{y'}(b, y_0(b), y_0'(b))\, Y(b).$$

Of most interest in these paragraphs are those admissible functions $y = y_0 + \alpha Y$ for which $Y(b) \neq 0$, for otherwise $y = y_0 + \alpha Y$ would be an admissible function for the fixed endpoint problem belonging to y_0. As a consequence, we must have $f_{y'}(b, y_0(b), y_0'(b)) = 0$. This concludes the proof of the following theorem.

(7.30) Theorem

If y_0 is a weak solution of the variational problem with free right endpoint, then y_0 fulfills the natural boundary condition

$$f_{y'}(b, y_0(b), y_0'(b)) = 0. \tag{7.4}$$

The words *'natural boundary condition'* can be explained in the following way: In solving the problem with a free right endpoint, the choice of that right endpoint will be given to the 'interplay of forces'. If we compare the minimum value m of the variational problem with the minimal values $m(y_b)$ of the different fixed boundary problems (which depend on the ordinate y_b of the right endpoint), we see that $m \leq m(y_b)$. If, therefore, a definite boundary condition – not the natural boundary condition – is to be found, some constraint on the endpoint must be exerted. In physical problems this is usually in the form of a force.

(7.5) Example

Among all the piecewise smooth curves in the xy-plane, find that curve $y = y(x)$ which joins the points $A(0,0)$ with a point B on the line $x = b > 0$ in such a way that the length of the curve is smallest. The problem is compactly stated by

$$\int_a^b \sqrt{1 + y'^2(x)}\, dx \to \min, \quad y(a) = 0, \quad y(b) \text{ arbitrary}.$$

The natural boundary condition requires $y_0'(b) = 0$. So the only extrema possible for the solution is $y_0 \equiv 0$.

(7.6) Example: The Ramsey growth model
The attentive reader of Section 2.3 will have noted that the question of boundary conditions for the desired function $K(t)$ remained open and that only the variational integral

$$J(K) = \int_0^T a \left(bK(t) - \frac{dk}{dt}(t) - C^* \right)^2 dt$$

was given. Here a, b and C^* are positive constants. The capital stock $K(0)$ at the initial time $t = 0$ of the planning period is assumed to be known: $K(0) = K_0$; on the other hand, the planner will not want to prescribe how large the capital will be at time $t = T$, for he is required only to maximize the total profit. We have therefore, a variational problem with free right endpoint before us. We discussed the extremals in Section 2.3; there can be no corners. The natural boundary conditions require that

$$bK(T) - \frac{dK}{dt}(T) - C^* = 0.$$

Consequently, the integration constant c_1 must be chosen equal to 0 so that the only extremals in question are the extremals

$$K(t) = \frac{C^*}{b} + \left(K_0 - \frac{C^*}{b} \right) e^{bt}.$$

(7.7) Exercise
Find functions y which satisfy the necessary conditions for a weak solution of the following problem: Among all of the piecewise smooth curves on the sphere which join the point $A = (0, 0)$ with a point B on the meridian $x = b \leq \pi$, find the curve which has the smallest length. (See Section 2.3 for the variational integral and technical particulars.)

After variational problems with free right endpoints have been discussed throughly, we should at least give results for variational problems with free left endpoints or simply with free endpoints (right and/or left).

(7.8) Theorem

Suppose y_0 is a weak solution of the following variational problem: From among all piecewise smooth functions $y : [a, b] \mapsto \mathbb{R}$ for which

$$y(b) = y_b$$

$$\text{is given, } y(a) \text{ is arbitrary (free left endpoint),} \tag{7.9}$$

or

$$y(a) \text{ and } y(b) \text{ are arbitrary (free endpoints),} \tag{7.10}$$

find that one which gives a minimum value to the variational integral

$$J(y) = \int_a^b f(x, y(x), y'(x))\, dx.$$

Then y_0 satisfies, respectively, the following natural boundary conditions:

$$f_{y'}(a, y_0(a), y_0'(a)) = 0 \quad \text{(free left endpoint)} \tag{7.11}$$

or

$$f_{y'}(a, y_0(a), y_0'(a)) = 0 = f_{y'}(b, y_0(b), y_0'(b)) \quad \text{(both endpoints free).} \tag{7.12}$$

The conclusion (7.12) is best proved by using two families of admissible functions $y = y_0 + \alpha Y$; for the first family it is assumed that $Y(a) = 0, Y(b) \neq 0$, while for the second $Y(a) \neq 0$ and $Y(b) = 0$.

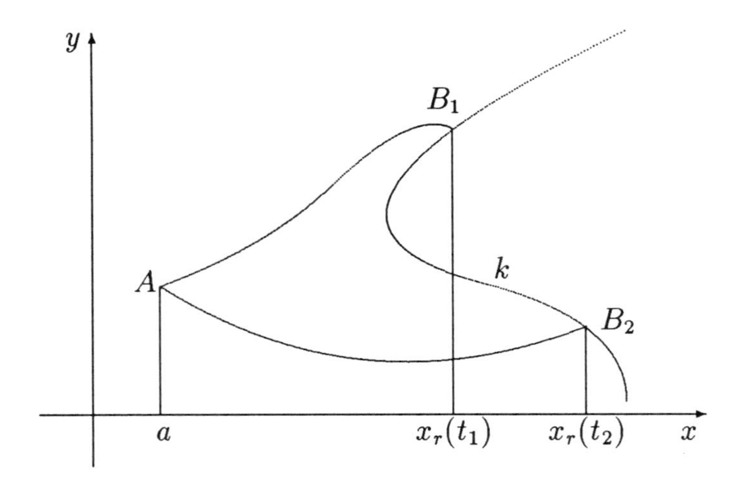

Fig. 7.1. Right endpoint lying on a curve.

7.2 VARIATIONAL PROBLEMS BETWEEN A POINT AND A CURVE

Here we will deal with variational problems for which the left endpoint is a fixed point A while the right endpoint lies on a given curve k.

The problem is, in principle, no different from the problem with free endpoints. The details, however, are somewhat more complicated. This is because, in general, the domains of definition of the different admissible functions are themselves different (cf. Fig. 7.1).

In order to represent the *boundary curve* k here, it is convenient to use a parametric representation, i.e. an invertible one-to-one (bijective) transformation of the t-interval I_t to points of the curve k;

$$t \in I_t \mapsto (x_r(t), y_r(t)),$$

where x_r and y_r are smooth functions on I_t for which it holds that

$$x_r'^2(t) + y_r'^2(t) \neq 0 \quad \text{for all} \quad t \in I_t.$$

To derive the necessary boundary conditions, we assume that the solution $y_0 : [a,b] \to \mathbb{R}$ of the problem is piecewise smooth and that it has corners at the points $x = a_i$. The function y_0 belongs to a family of admissible functions $y(\cdot,\alpha)$, $|\alpha| < \alpha_0$, where $y_0(x) = y(x,0)$ for all $x \in [a,b]$. We require of the function $y(\cdot,\alpha)$ that its partial derivative with respect to α be continuous and that the mixed derivative $\frac{\partial^2 y}{\partial x \partial \alpha}$ be piecewise continuous. The right endpoint of the 'curve' $y(\cdot,\alpha)$ (α fixed) lies on the boundary curve k so that if we employ the curve parameter $t(\alpha)$, then the domain of definition of $y(\cdot,\alpha)$ is $[a, x_r(t(\alpha))]$. (It is to be noted that the specification $y(x,\alpha) = y_0(x) + \alpha Y(x)$ is no longer meaningful here.) We now investigate the function $\Phi(\alpha) = J(y(\cdot,\alpha))$, which has a relative minimum at the point $\alpha = 0$

$$\Phi(\alpha) = \int_a^{x_r(t(\alpha))} f(x, y(x,\alpha), y_x(x,\alpha))\, dx. \tag{7.13}$$

When we differentiate we must remember that α appears in the upper limit of the integral

$$\Phi'(0) = f(b, y_0(b), y_0'(b)) \frac{dx_r}{dt}\frac{dt}{d\alpha}(0) + \psi = 0,$$

where

$$\psi = \int_a^b \left[f_y^0(x) \frac{\partial y}{\partial \alpha}(x,0) + f_{y'}^0(x) \frac{\partial^2 y}{\partial \alpha \partial x}(x,0) \right] dx.$$

Integration by parts leads to

$$\begin{aligned}\psi = &\int_a^b \left[f_y^0(x) - \frac{d}{dx} f_{y'}^0(x) \right] \frac{\partial y}{\partial \alpha}(x,0) dx + f_{y'}^0(b) \frac{\partial y}{\partial \alpha}(b,0) \\ &+ \sum_i [f_{y'}(a_i, y_0(a_i), y_{0l}'(a_i)) \\ &- f_{y'}(a_i, y_0(a_i), y_{0r}'(a_i))] \frac{\partial y}{\partial \alpha}(a_i, 0).\end{aligned}$$

Since y_0 satisfies both the Euler equation and the corner conditions, the above reduces to

$$\psi = f^0_{y'}(b)\frac{\partial y}{\partial \alpha}(b,0)$$

so that

$$\begin{aligned}\Phi'(0) = f(b, y_0(b), y_0'(b))&\frac{dx_r}{dt}(t(0))\frac{dt}{d\alpha}(0)\\ &+ f^0_{y'}(b)\frac{\partial y}{\partial \alpha}(b,0).\end{aligned} \tag{7.14}$$

A comparison of the natural boundary conditions (7.4) with these last results shows that b and $Y(b)$ in (7.4) correspond, respectively, to x_r and $\frac{\partial y}{\partial \alpha}(b,0)$. In the condition stated in (7.14) the factor $\frac{\partial y}{\partial \alpha}(b,0)$ describes the change of y with respect to α while $x = b$ is held fixed. The change in y along the boundary curve k plays a larger role here. And this becomes apparent if we compare the two representations of the value of y at the right endpoint. The value of $y(\cdot, \alpha)$ at the right endpoint is also the ordinate of the corresponding point on the boundary curve k:

$$y(x_r(t(\alpha)), \alpha) = y_r(t(\alpha)). \tag{7.15}$$

If we differentiate (7.15) with respect to α we obtain, at the point $\alpha = 0$,

$$\frac{\partial y}{\partial x}(b,0)\frac{dx_r}{dt}(t(0))\frac{dt}{d\alpha}(0) + \frac{\partial y}{\partial \alpha}(b,0) = \frac{dy_r}{dt}(t(0))\frac{dt}{d\alpha}(0).$$

Solving for $\frac{\partial y}{\partial \alpha}(b,0)$ and putting the resulting expression into (7.14) gives

$$\begin{aligned}\{f^0_{y'}(b)y_r'(t(0)) + [f(b, y_0(b), y_0'(b))&\\ - f^0_{y'}(b)y_0'(b)] \cdot x_r'(t(0))\}&\frac{dt}{d\alpha}(0) = 0.\end{aligned}$$

Since there exist families of curves for which $\frac{dt}{d\alpha}(0) \neq 0$, it must be that the bracketed expression vanishes, $\{\cdots\} = 0$.

(7.16) Theorem

Let y_0 be a solution of the variational problem for which the left endpoint of an admissible function is held fixed and the right end lies on a curve k with the parametric representation $t \to (x_r(t), y_r(t))$. Then y_0 satisfies the transversality conditions

$$f^0_{y'}(b)y'_r(t_0) + [f(b, y_0(b), y'_0(b)) - f^0_{y'}(b)y'_0(b)]x'_r(t_0) = 0, \quad (7.17)$$

where t_0 is the parameter of the right endpoint of y_0.

An analogous result holds in the case in which the right endpoint of an admissible function is held fixed while the left endpoint is free to move along a curve k. If conditions (7.17) are satisfied, we say that the curve k cuts the curve y_0 *transversally.*

Exercise

(7.18) Find the necessary boundary conditions for a variational problem in which the left endpoint is free to move on a curve k_ℓ and the right endpoint on a curve k_r.

(7.19) Show that for the variational problem of the shortest curve joining a point A and a curve k in the plane, then curve k cuts the extremal y_0 transversally precisely when k cuts the graph of y_0 orthogonally. Recall that the integrand f is given by

$$f(x, y, y') = \sqrt{1 + y'^2}.$$

7.3 SUFFICIENT CONDITIONS AND MOVABLE BOUNDARIES

While the necessary conditions for the fixed boundary problems are also necessary conditions for the problem with movable endpoints – only one additional necessary boundary condition is needed – the sufficient conditions of Chapter 3 hold for problems with movable boundary while the conditions of Chapter 6 are no longer needed. This comes about because, for the development of the sufficient conditions, the difference $J(y) - J(y_0)$, where y is an arbitrary admissible function and y_0 is the candidate for the solution of the problem, must be restricted. As we have already noted in Section 7.1, there are many more admissible functions for the problem with movable endpoints than for the problem with fixed endpoints. In spite of this, the ideas which the sufficient conditions of Chapters 3 and 6 contain can be modified so that they will achieve our purposes for problems with movable endponts. We see, therefore, how important it is to understand the fundamental ideas. In what follows we present a sufficient condition for convex integrands. There are several possibilities for the construction of sufficient conditions with the help of the invariant integral. We will derive one such sufficient condition.

So consider a variational problem with free right and/or left endpoints. Let the integrand f of the variational integral be convex in (y, y') in the sense of (3.13), (3.14). As in Section 3.2, we estimate the following difference:

$$\begin{aligned} J(y) - J(y_0) &= \int_a^b [f(x, y(x), y'(x)) - f(x, y_0(x), y_0'(x))]\, dx \\ &\geq \int_a^b [f_y^0(x)(y(x) - y_0(x)) \\ &\quad + f_{y'}^0(x)(y'(x) - y_0'(x))]\, dx \qquad (3.15) \\ &= \int_a^b \left[f_y^0(x) - \frac{d}{dx} f_{y'}^0(x) \right] (y(x) \\ &\quad - y_0(x))\, dx + [f_{y'}^0 (y - y_0)]_{x=a}^{x=b} + \text{`corner terms'}. \end{aligned}$$

The corner terms – if $y - y_0$ has corners – come into consideration because of the integration by parts (just as in similar earlier formulae). They vanish, however, because of the continuity of y. The integral on the right side of this inequality equals zero, because y_0 satisfies the Euler equation. The second sum on the right also vanishes, independent of which problem is being considered; if $x = a$ or $x = b$ is a fixed endpoint, then

$y(x) - y_0(x) = 0$; if, however, the boundary value is arbitrary at $x = a$ or $x = b$, then $f^0_{y'}(x) = 0$ in accordance with the natural boundary condition. We summarize these results in the following theorem.

(7.20) Theorem
Let y_0 be a smooth extremal and admissible function of the variational problem with free boundary (i.e. free endpoints) and suppose y_0 satisfies the natural boundary conditions. If the integrand satisfies the assumptions (3.13) *and* (3.14), *i.e. if it is convex in y and y', then y_0 is a solution of the variational problem.*

Example from Ramsey growth model
In Section 3.1 it was shown that the integrand of the variational integral is convex. The extremal which satisfies the natural boundary condition was determined in (7.6). According to Theorem (7.20), this extremal is a (uniquely determined) solution of the problem with free right endpoint.

Just as with the fixed boundary problem, so here with the help of extremal fields sufficient conditions can be set up. We will do this for the variational problem between a point and a curve. Thus, suppose y_0 is an extremal which satisfies the boundary condition $y_0(a) = y_a$ and also the transversality condition (7.17). Let the boundary curve k have the parametric representation $(x_r(t), y_r(t))$. We assume also that there is a field of extremals $\tilde{y}$ in which y_0 is embedded and that all extremals of the extremal field satisfy the transversality conditions. This assumption will be discussed at greater length below.

The associated field will be denoted by G and the slope function by p. We assume that G is open, connected and simply connected. Then there is an independence integral I^* for the field of extremals,

$$I^* = \int \{[f(x, y, p(x, y)) - p(x, y) f_{y'}(x, y, p(x, y))]\, dx$$
$$+ f_{y'}(x, y, p(x, y))\, dy\} \quad \text{(cf. (6.10))}.$$

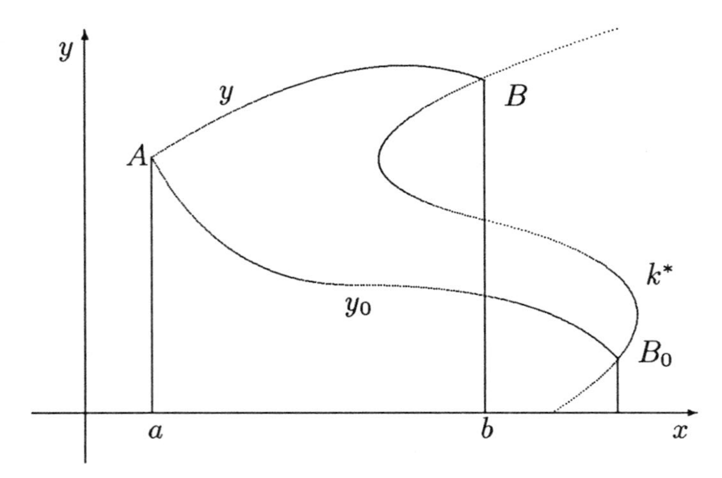

Fig. 7.2. Boundary curve k^*.

If I^* is evaluated for a piece of the boundary curve k (i.e. if dx is replaced by $x_r'(t)\,dt$, dy by $y_r'(t)\,dt$ and x and y, respectively, by $x_r(t)$ and $y_r(t)$), then because of the transversality conditions (7.17) the integrand is identically zero. If now y is an admissible function whose graph lies in the field G, then $I^*(y) = I^*(y_0) + I^*(K^*) = I^*(y_0)$, where k^* is that part of the boundary curve from the right endpoint B_0 of y_0 to the right endpoint $B = (b, y(b))$ of y (cf. Fig. 7.2).

We may now write the difference

$$J(y) - J(y_0) = J(y) - I^*(y_0) = J(y) - I^*(y)$$
$$= \int_a^b \mathcal{E}(x, y(x), p(x, y(x)), y'(x))\,dx$$

($\mathcal{E}$ is the Weierstrass excess function). Thus we have

(7.21) Theorem

Let y_0 be an extremal of the variational problem between the point $A = (a, y_a)$ and the curve k. Let y_0 satisfy the boundary conditions $y_0(a) = y_a$ and the transversality condition. If there is a field of extremals with the properties:

(i) *y_0 is embedded in $\tilde{y}$;*

(ii) *all extremals of the field of extremals $\tilde{y}$ are cut transversally by the boundary curve k;*

(iii) *the field G of $\tilde{y}$ is open, connected and simply connected, and if for all $(x, y) \in G$ it holds that*

$$\mathcal{E}(x, y, p(x, y, q)) \geq 0 \quad \text{for all} \quad q \in \mathbb{R}$$

$$(p(x, y) \text{ denotes the slope of } \tilde{y} \text{ at } (x, y)),$$

then y_0 is a strong solution of the variational problem. Furthermore, $J(y) \geq J(y_0)$ for all admissible functions y whose graph lies in the field G.

Example

To solve the following variational problem no calculus of variations is needed. It is, nonetheless, an appropriate example to make vivid the concepts and phenomenon which can appear.

Among all piecewise smooth 'curves' y which join the point $A = (a, 0)$ with a point P on the semi-circle k_r having radius r,

$$k_r \hat{=} \Big\{(x, y) \in \mathbb{R}^2; x = r\cos t = 8x_r(t),$$
$$y = r\sin t = y_r(t), -\frac{\pi}{2} < t < \frac{\pi}{2}\Big\},$$

find the 'curve' having the shortest length. The extremals of this variational problem are pieces of straight lines. A curve cuts another transversally if it intersects the other orthogonally (cf. Exercise (7.19)). We will discuss the variational problem for different initial points A.

1. If $0 < a < r$, then the null function $y_0 = 0$ satisfies all the necessary conditions. Theorem (7.21) is described in Fig. 7.3. Thus y_0 is the solution of the variational problem.

2. If $a = 0$, then all line segments $y(x) = \alpha x$ satisfy the transversality conditions. Theorem (7.21) is not applicable since $A = (0, 0)$ is not in G. It can be shown that all of these line segments are solutions of the variational problem.

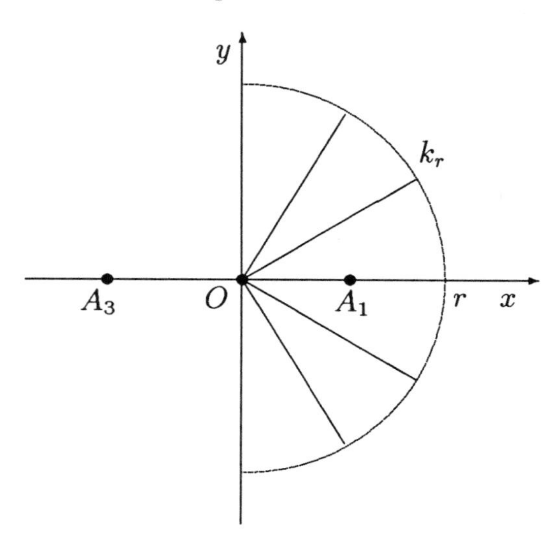

Fig. 7.3. Shortest curve joining the point A to the semi-circle k_r.

3. If $a < 0$, then $y_0 = 0$ is the only extremal which satisfies the transversality conditions (7.17); but in this case the sufficient condition of Theorem (7.21) is not satisfied. This is no surprise. Here the variational problem has no solution since it can be shown that the distance $d(t)$ from the point A to the point $P(t), P(t) = (r\cos t, r\sin t),\ -\frac{\pi}{2} < t < \frac{\pi}{2}$,

$$d(t) = \sqrt{r^2 \sin^2 t + (r\cos t - a)^2}$$

has no minimum in $(-\frac{\pi}{2}, \frac{\pi}{2})$. The points $(0, +r)$ and $(0, -r)$ do not belong to the curve k_r.

We conclude with a discussion of the assumption of Theorem (7.21), i.e. that there exists a field of extremals which is cut transversally by k. Under the assumption (regularity condition)

$$f_{y'y'}(x_r(t), y_r(t), q) \neq 0 \quad \text{for all} \quad t \text{ and } q \text{ in } \mathbb{R} \tag{7.22}$$

the extremals can be found as solutions of an initial value problem (cf. (2.10)). So if we take the initial point of the extremal to be the point $(x_r(t), y_r(t))$ on the boundary curve, then the initial direction $p(t)$ is determined by the transversality condition (7.17):

$$\begin{aligned}\psi(t,p(t)) &= f_{y'}(x_r(t),y_r(t),p(t))y_r'(t)\\ &\quad+[f(x_r(t),y_r(t),p(t))\\ &\quad-f_{y'}(x_r(t),y_r(t),p(t))p(t)]x_r'(t)\\ &= 0.\end{aligned}$$

From the implicit function theorem this equation may be solved (locally, in a neighbourhood of $p(t_0) = y_0'(b)$) for $p(t)$ if

$$\frac{\partial\psi}{\partial p}(t_0, y_0'(b)) \neq 0.$$

This condition is satisfied if, in addition to (7.22), it also holds that

$$f(b, y_0(b), y_0'(b)) \neq 0. \tag{7.23}$$

Thus, under the assumptions (7.22) and (7.23), through each point of k which is sufficiently near $(b, y_0(b))$, there passes an extremal $y(\cdot, t)$ which is cut transversally by k. The curve y_0 is one of these extremals. Whether these extremals form an appropriate field of extremals over a sufficiently large part of their domain of definition and whether y_0 is embedded in this field must be determined for individual cases.

Exercises

(7.24) Among all piecewise smooth functions $y : [0,1] \to \mathbb{R}$ with $y(0) = 0$, find those for which the intergral

$$J(y) = \int_0^1 (y^2(x) + y'^2(x) + 2y(x)e^x)\,dx$$

has a minimal value (cf. Exercise (6.28)). This is a problem with free right endpoint. Check both sufficient conditions.

(7.25) Among all piecewise smooth functions $y : [0,1] \mapsto \mathbb{R}$ with $y(0) = 0$, $y(x) > 0$ for $x \in (0,1), y(1)$ arbitrary, find those for which the integral

$$J(y) = \int_0^1 -y(x)\sqrt{1 - y'^2(x)}\,dx$$

has its smallest value (cf. Example (1.4)).

8

Variational problems depending on several functions

Frequently in mathematical applications more than a single function y is needed. We need only think of the path of a point mass in space: For each point t in time the position of the point mass is described by its coordinates $x = x(t)$, $y = y(t)$ and $z = z(t)$, i.e. by three functions. Another example is given by the generalization of the Ramsey growth model in which optimization involving the production of two or more products is to be worked through (in a yet to be defined sense).

Thus we need to describe the process by means of finitely many functions of a single variable, i.e. $y_1(x), y_2(x)$, $y_3(x) \dots y_n(x)$. We will combine these functions in a single vector function $\underline{y}$:

$$\underline{y}(x) = (y_1(x), y_2(x), \dots, y_n(x)).$$

(In what follows, the underline $\underline{\quad}$ will indicate that a vector function is being discussed.) We review briefly the calculation rules for vector functions.

Vector functions are added and multiplied with scalars componentwise:

$$\underline{y}(x) + \underline{z}(x) = (y_1(x) + z_1(x), \dots, y_n(x) + z_n(x)),$$

$$c\underline{y}(x) = (cy_1(x), \dots, cy_n(x)), \quad c \in \mathbb{R}.$$

We will write the scalar product and the norm for vector functions in the usual way

$$\underline{y}(x) \cdot \underline{z}(x) = y_1(x)z_1(x) + \cdots + y_n(x)z_x = \sum_{i=1}^{n} y_i(x)z_i(x),$$

$$|\underline{y}(x)| = \sqrt{\underline{y}(x) \cdot \underline{y}(x)}.$$

A vector function $\underline{y} : [a,b] \to \mathbb{R}^n$ is continuous, r-fold smooth or piecewise smooth if all of its components are continuous, r-fold smooth or piecewise smooth, respectively. The derivative of a vector function $\underline{y}$ is also calculated componentwise

$$\underline{y}'(x) = (y_1'(x), y_2'(x), \ldots, y_n'(x)).$$

The distances d_0 and d_1 of the vector function $\underline{y}$ from $\underline{z}$ can be defined just as in the case of the single function

$$d_0(\underline{y}, \underline{z}) = \max_{x \in [a,b]} |\underline{y}(x) - \underline{z}(x)|,$$

$$d_1(\underline{y}, \underline{z}) = \sup_{x \in [a,b]} [|\underline{y}(x) - \underline{z}(x)| + |\underline{y}'(x) - \underline{z}'(x)|].$$

For the variational problem depending on several functions y_i, $i = 1, 2, \ldots, n$, all of which are combined in the single vector function $\underline{y}$, we make the following assumptions with respect to the integrand f of the variational integral: Assume f to be defined on an open set M in the $(x, y_1, y_2, \ldots, y_n, y_1', y_2', \ldots, y_n')$ space (briefly, the xy_iy_i'-space) and to have continuous first, second and third partial derivatives with respect to all variables there. Here the y_i' are also variables for which the derivatives y_i' can be substituted. The variational integral is now

$$\begin{aligned} J(y) &= \int_a^b f(x, y_1(x), \ldots, y_n(x), y_1'(x), \ldots, y_n'(x))\, dx \\ &= \int_a^b f(x, \underline{y}(x), \underline{y}'(x))\, dx. \end{aligned}$$

Admissible functions are all vector functions $\underline{y}$ of x which are piecewise smooth on $[a,b]$ and which satisfy certain boundary conditions. We will not determine the kind of boundary conditions (fixed, free, etc.) here.

A simple example is expressed by the problem, find the shortest curve joining the point $A = (a, y_a, z_a)$ to the point $B = (b, y_b, z_b)$ in space. If the problem is restricted to piecewise smooth curves which can be expressed by two functions y and z of x, then the variational problem has the form

$$J(y, z) = \int_a^b \sqrt{1 + y'^2(x) + z'^2(x)}\, dx \to \min.$$

$$\text{Boundary conditions:} \quad y(a) = y_a, \quad y(b) = y_b,$$
$$z(a) = z_a, \quad z(b) = z_b.$$

(We may assume here without loss of generality that $a < b$.)

8.1 THE CONDITIONS OF THE FIRST VARIATION

Variational problems which depend on several functions are treated in the same way as the problems of one function. We will show this by deriving the necessary conditions of the first variation for the fixed endpoint problem and for the problem with free right endpoint.

As in Chapter 2, we assume that $\underline{y}_0$ is a weak solution of the variational problem; $\underline{y}_0$ is therefore an admissible function and there is a $\delta > 0$ such that $J(\underline{y}_0) \leq J(\underline{y})$ for all admissible functions $\underline{y}$ with $d_1(\underline{y}, \underline{y}_0) < \delta$.

If $\underline{Y}$ is a piecewise smooth vector function which satisfies the boundary conditions

$$\underline{Y}(a) = \underline{Y}(b) = 0 \quad \text{(fixed endpoint problem)}$$

or

$$\underline{Y}(a) = 0,$$

$Y(b)$ arbitrary (problem with free right endpoint),

then there is an $\alpha_0 > 0$ such that for each fixed α with $|\alpha| < \alpha_0$, the vector function

$$\underline{y}(x,\alpha) = \underline{y}_0(x) + \alpha\underline{Y}(x)$$

is admissible with $d_1(\underline{y}(\cdot,\alpha),\underline{y}_0) < \delta$. It follows that for $|\alpha| < \alpha_0$,

$$\Phi(\alpha) = J(\underline{y}(\cdot,\alpha)) \geq J(\underline{y}_0).$$

For what follows it is enough that we only consider functions $\underline{Y}$ for which only one component is different from zero:

$$\underline{Y}(x) = (0,\ldots,0,Y_i(x),0,\ldots,0), \quad i \in \{1,2,\ldots,n\}. \tag{8.1}$$

Then if at the point $\alpha = 0$ the function Φ has a relative minimum, it must be that $\Phi'(0) = 0$

$$\Phi'(0) = \int_a^b [f^0_{y_i}(x)Y_i(x) + f^0_{y'_i}(x)Y'_i(x)]\,dx = 0,$$

where we have been using the abbreviating notations

$$\begin{aligned} f^0_{y_i}(x) &= f_{y_i}(x,\underline{y}_0,\underline{y}'_0(x)), \\ f^0_{y'_i}(x) &= f_{y_i}(x,\underline{y}_0(x),\underline{y}'_0(x)). \end{aligned} \tag{8.2}$$

If we apply integration by parts to the first term in the integrand for $\Phi'(0)$ we obtain

$$\begin{aligned} \Phi'(0) &= \int_a^b \left[f^0_{y'_i}(x) - \int_a^x f^0_{y_i}(u)\,du\right] Y_i(x)\,dx \\ &\quad + \left[\int_a^x f^0_{y_i}(u)\,du\,Y_i(x)\right]_{x=a'}^{x=b}. \end{aligned} \tag{8.3}$$

For the fixed boundary problem, the second term on the right side of (8.3) is equal to 0. From the Lemma of du Bois–Reymond we can further conclude that

$$f^0_{y'_i}(x) - \int_a^x f^0_{y_i}(u)\,du = c = \text{ constant for } x \in [a,b]. \tag{8.4}$$

Since i was an arbitrarily chosen index in (8.3), the relation (8.4) is valid for all $i = 1, 2, \ldots, n$. At the point $x \in [a, b]$ where $\underline{y}_0'$ is continuous, (8.4) can be differentiated with respect to x

$$\frac{d}{dx} f_{y_i'}^0(x) - f_{y_i}^0(x) = 0 \quad (i = 1, 2, \ldots, n). \tag{8.5}$$

These are the Euler differential equations for the variational problem. If, however, $\underline{y}_0$ has a corner at the point x_0, i.e. if the left- and right-hand derivatives $y_{0l}'(x)$ and $y_{0r}'(x)$ do not agree, then the Erdmann corner conditions follow from (8.4). That is,

$$\begin{gathered} f_{y_i'}(x_0, \underline{y}_0(x_0), \underline{y}_{0l}'(x_0)) = f_{y_i}'(x_0, \underline{y}_0(x_0), y_{0r}'(x_0)), \\ i = 1, 2, \ldots, n. \end{gathered} \tag{8.6}$$

A solution $\underline{y}_0$ of the variational problem with free right endpoints satisfies the conditions (8.4), since it is also a solution of the fixed endpoint problem with the boundary conditions $\underline{y}(a) = \underline{y}_0(a)$, $\underline{y}(b) = \underline{y}_0(b)$. So it follows from (8.3) and (8.4) that

$$\begin{aligned} \Phi'(0) = 0 &= \int_a^b cY_i'(x)\,dx + \int_a^b f_{y_i}^0(u)\,du\,Y_i(b) \\ &= c(Y_i(b) - Y_i(a)) + \int_a^b f_{y_i}^0(u)\,du\,Y_i(b) \\ &= (c + \int_a^b f_{y_i}^0(u)\,du)\,Y_i(b) - cY_i(a). \end{aligned}$$

But $Y_i(a) = 0$ and $Y_i(b)$ is arbitrary so that

$$0 = c + \int_a^b f_{y_i}^0(u)\,du = f_{y_i'}^0(b), \quad i = 1, \ldots, n. \tag{8.7}$$

These are the *natural boundary conditions.*

The Euler differential equations (8.5) are a system of n (in general, non-linear) differential equations of second order for the functions y_i. For if $\underline{y}_0$ is twice differentiable, then differentiation with respect to x can be carried out by application of the chain rule. Also, under the assumption that the determinant of the matrix having entries

$$f_{y_i'y_j'}(x, \underline{y}_0(x), \underline{y}_0'(x)), \quad i, j = 1, 2, \ldots, n,$$

is different from zero, it is possible to solve equation (8.5) for y''_{0i}. Thus the Euler differential equations form a system of n differential equations of second order for the desired functions y_i.

We can apply here the methods of Section 2.2 to solve the Euler differential equations under certain special assumptions. If f is not dependent on x and if $\underline{y}_0$ is a twice differentiable extremal, then the following function H must be constant:

$$H(x) = f(x, \underline{y}_0(x), \underline{y}'_0(x)) - \sum_{i=1}^{n} y'_{0i}(x) f^0_{y'_i}(x). \tag{8.8}$$

It can be shown easily that $H' = 0$.

Example

$$J(y) = \int_0^1 [y'^2_1(x) + (y'_2(x) - 1)^2 + y_1^2(x) + y_1(x) y_2(x)]\, dx \to \min .$$

Since the integrand is a quadratic function in (y_1, y_2, y'_1, y'_2), the Euler differential equations are linear

$$y''_1 - y_1 - \tfrac{1}{2} y_2 = 0,$$
$$y''_2 - \tfrac{1}{2} y_1 = 0.$$

The second equation can be solved for y_1 and the resulting expression for y_1 substituted in the first equation

$$y_1 = 2y''_2, \quad 2y_2^{iv} - 2y''_2 - \tfrac{1}{2} y_2 = 0$$

(y_2^{iv} denotes the fourth derivative of y_2). The linear homogeneous differential equation of fourth order in y_2 can be solved, as is well known, by setting $y_2(x) = e^{kx}$, where k is a complex number. Putting this in the differential equation yields values for k

$$k_{1,2} = \pm \sqrt{\frac{1}{2} + \sqrt{\frac{1}{2}}} \in \mathbb{R};$$

$$k_{3,4} = \pm \sqrt{\frac{1}{2} - \sqrt{\frac{1}{2}}} = \pm\, im \quad (i^2 = -1).$$

Thus the general solution of $y = (y_1, y_2)$ of the Euler differential equation is given by

$$\begin{aligned} y_1(x) &= 2C_1k_1^2e^{k_1x} + 2c_2k_2^2e^{k_2x} \\ &\quad -2C_3m^2\cos mx - 2C_4m^2\sin mx, \\ y_2(x) &= C_1e^{k_1x} + C_2e^{k_2x} + C_3\cos mx + C_4\sin mx, \end{aligned}$$

where $C_1, C_2, \ldots, C_4$ are constants of integration, i.e. arbitrary constants. We could also write

$$\begin{aligned} y_1(x) &= 2Ak_1^2\sinh(k_1x + B) - 2Cm^2\sin(mx + D), \\ y_2(x) &= A\sinh(k_1x + B) + C\sin(mx + D), \end{aligned}$$

with constants of integration A, B, C, D. For the fixed boundary problem extremals are sought which satisfy the prescribed boundary conditions

$$y_1(0) = y_{10}, \quad y_1(1) = y_{11},$$

$$y_2(0) = y_{20}, \quad y_2(1) = y_{21}.$$

From these we can obtain a linear system of equations in the four unknowns C_1, C_2, C_3 and C_4, which has a unique solution.

In the case of the problem with free right endpoints the natural boundary condition for the extremal $\underline{y}_0 = (y_{01}, y_{02})$ is

$$\begin{aligned} f^0_{y_1'}(1) &= 2y_{01}' = 0 \quad \text{or} \quad y_{01}'(1) = 0, \\ f^0_{y_2'}(1) &= 2(y_{02}'(1) - 1) = 0 \quad \text{or} \quad y_{02}'(1) = 1. \end{aligned}$$

These results, combined with the initial conditions $y_{01}(0) = y_{10}$, $y_{02}(0) = y_{20}$, provide four linear conditions from which the values C_1, C_2, C_3 and C_4 may be uniquely determined.

8.2 FURTHER NECESSARY AND SUFFICIENT CONDITIONS

The remaining necessary and sufficient conditions we have thus far discussed may be generalized to variational problems involving several unknown functions. In each instance, the basic idea is the same as that employed in connection with problems involving a single unknown function.

We will only provide the results.

(8.9) Theorem

(Sufficient conditions when f is a convex function.) Let the variational problem have fixed endpoints or one or two free endpoints. If the integrand f of the variational integral is a convex function of the variables $(y_1, y_2, \ldots, y_n, y_1', y_2', \ldots, y_n')$ for each $x \in [a, b]$ (cf. Definition (3.2)),then every admissible, smooth vector function which satisfies the Euler differential equations is a solution of the problem.

(8.10) Theorem

Let y_0 be a weak solution of the variational problem. Then the following necessary condition of Legendre holds: For all x values for which $\underline{y}_0$ is differentiable and for all vectors $u = (u_i) \in \mathbb{R}^n$,

$$\sum_{i,j=1}^{n} f_{y_i' y_j'}(x, \underline{y}_0(x), \underline{y}_0'(x)) u_i u_j \geq 0. \tag{8.11}$$

In case $\underline{y}_0$ is also a strong solution of the variational problem, then the Weierstrass necessary condition holds: For all x for which $\underline{y}_0$ is differentiable and for all $q \in \mathbb{R}^n$,

$$\mathcal{E}(x, \underline{y}_0(x), \underline{y}_0', q) \geq 0, \tag{8.12}$$

where the Weierstrass excess function $\mathcal{E}$ is now defined by

$$\begin{aligned}\mathcal{E}(x, y_j, y_j', q_j) = & f(x, y_j, q_j) - f(x, y_j, y_j') \\ & - \sum_{i=1}^{n} f_{y_i'}(x, y_j, y_j')(q_i - y_i').\end{aligned} \tag{8.13}$$

Remark

The condition (8.11) is satisfied precisely when all eigenvalues of the symmetric matrix with elements

$$f_{y_i'y_j'}(x, \underline{y}_0, \underline{y}_0'(x)) \quad (i, j = 1, 2, \ldots, n)$$

are equal to or greater than zero. The extremal $\underline{y}_0$ is said to be *regular* if the determinant of this matrix is different from 0 for all $x \in [a, b]$.

The second derivative of the function $\Phi(\alpha) = J(\underline{y}_0 + \alpha \underline{Y})$ at the point $\alpha = 0$ leads to the second variation and the accessory variational problem (cf. Section 5.1):

$$\Phi''(0) = \delta^2 J = \int_a^b 2\Omega(x, \underline{Y}(x), \underline{Y}'(x))\, dx = I(\underline{Y}),$$

where

$$\begin{aligned}2\Omega(x, Y_j, Y_j') &= \sum_{i,j=1}^{n} [f^0_{y_i y_j}(x) Y_i Y_j + 2 f^0_{y_i y_j}(x) Y_i Y_j' + f^0_{y_i' y_j'}(x) Y_i' Y_j'], \\ f^0_{y_i y_j}(x) &= f_{y_i y_j}(x, \underline{y}_0(x), \underline{y}_0'(x)), \\ f^0_{y_i y_j'}(x) &= f_{y_i y_j'}(x, \underline{y}_0(x), \underline{y}_0'(x)), \\ f^0_{y_i' y_j'}(x) &= f_{y_i' y_j'}(x, \underline{y}_0(x), \underline{y}_0'(x)).\end{aligned}$$

The Euler differential equations for the accessory variational problem $I(Y) \to \min$ are called the *Jacobi equations.*

A point $x = c$ is called a *conjugate point* of $x = a$ if there is a two-fold smooth solution $\underline{Y}$ of the Jacobi equations with the initial value $Y(a) = 0$ and such that Y is not identically 0 on any interval of positive length, but for which $Y(c) = 0$.

(8.14) Theorem (Jacobi's necessary condition)
Let $\underline{y}_0$ be a four-fold smooth, regular, weak or strong solution of the variational problem. Then the interval (a, b) contains no point conjugate to $x = a$.

We come next to the sufficient condition which is associated with a path-independent integral. The integral I^* will be defined with the help of a field of extremals and in a way similar to that used for the variational problem with one unknown function (cf. Definition (6.8)). The only difference consists in this: That the field of extremals must have the additional property that it be a central field. The following definition of a central field is a generalization of Definition (6.7).

(8.15) Definition
A *central field* is a vector function defined on a subset of the $(x, \alpha_1, \alpha_2, \dots, \alpha_n)$-space, which has smooth second partials and, additionally, has the following properties:

(i) for each fixed n-tuple $(\alpha_1, \alpha_2, \dots, \alpha_n)$, the vector function $\underline{\tilde{y}}(\cdot, \alpha_1, \alpha_2, \dots, \alpha_n)$ satisfies the Euler differential equations. There is a point (x_0, y_{i0}) (called the centre) such that for all $\alpha_1, \alpha_2, \dots, \alpha_n$, $\tilde{y}_i(x_0, \alpha_1, \alpha_2, \dots, \alpha_n) = y_{i0}$, $i = 1, 2, \dots, n$.

There is an open non-empty subset M of the domain of definition of $\underline{\tilde{y}}$ for which

(ii) for all $(x, \alpha_1, \alpha_2, \dots, \alpha_n) \in M$ the functional determinant

$$\det \left(\frac{\partial \tilde{y}_i}{\partial \alpha_j}(x, \alpha_1, \alpha_2, \dots, \alpha_n) \right)_{i,j=1,2,\cdots,n}$$

is not equal to 0; and

(iii) through each point of the graph $G = \{(x, \underline{\tilde{y}}(x, \alpha_1, \alpha_2, \dots, \alpha_n) \in \mathbb{R}^{n+1} : (x, \alpha_1, \dots, \alpha_n) \in M\}$ – called the field – only one extremal of the field of extremals passes. That is, for each $(x, y_i) \in G$ there is a unique set of parameter values $\alpha_1, \alpha_2, \ldots, \alpha_n$ for which $y_i = \tilde{y}_i(x, \alpha_1, \alpha_2, \ldots, \alpha_n)$, $i = 1, 2, \ldots, n$.

The one-to-one correspondence defined in (iii) will be denoted by $(x, y_i) \mapsto \underline{\hat{\alpha}}(x, y_i)$. Thus for a central field, the *slope function* $\underline{p}$ on M is defined by

$$\underline{p}(x, y_i) = \tilde{\underline{y}}_x(x, \underline{\hat{\alpha}}(x, y_i)).$$

The vector function $\underline{p}$ is the direction vector of the extremals through the point $(x, y_i) \in G$. An open and connected set G in $\mathbb{R}^m$ is said to be *simply connected* if every closed curve k within G can be continuously deformed to a 'point curve'; i.e. if for each closed curve k in G there is a continuous transformation $\underline{H} : [0,1] \times [0,1] \mapsto G$ so that $\underline{H}(\cdot, 0) : [0,1] \mapsto G$ is a parametric representation of k, $\underline{H}(t, 1) = \text{constant} = P$ for all $t \in [0,1]$, and $\underline{H}(0, \alpha) = \underline{H}(1, \alpha)$ for all $\alpha \in [0,1]$. ($\underline{H}$ describes a family of closed curves in G; α is the parameter of the family.)

(8.16) Theorem

Let $\tilde{\underline{y}}$ be a central field with slope function $\underline{p}$ and suppose the field G of $\tilde{\underline{y}}$ is open, connected and simply connected. Then

$$I^*(\underline{y}) = \int_{x_1}^{x_2} [A(x, \underline{y}(x)) + \sum_{i=1}^{n} B_i(x, \underline{y}(x)) y_i'(x)]\, dx,$$

where

$$A(x, y_j) = f(x, y_j, p_j(x, y_j)) - \sum_{i=1}^{n} p_i(x, y_j) f_{y_i'}(x, y_j, p_j(x, y_j))$$

and

$$B_i(x, y_j) = f_{y_i'}(x, y_j, p_j(x, y_j)),$$

is a path-independent integral; i.e. it has the same value for any two vector functions $\underline{y}$ and $\underline{y}^*$ which satisfy the same boundary conditions $\underline{y}(x_1) = \underline{y}^*(x_1)$, $\underline{y}(x_2) = \underline{y}^*(x_2)$, and whose graphs lie in G.

The proof of this theorem proceeds essentially as the proof of Theorem (6.10), which represents the special case $n = 1$. The assumption that all extremals of the field of extremals pass through a single point (x_0, y_{i0}) will be needed in order to prove the integrability conditions for the system of partial differential equations for the desired potential function P:

$$\frac{\partial P}{\partial x}(x, y_j) = A(x, y_j), \quad \frac{\partial P}{\partial y_i}(x, y_j) = B_i(x, y_j).$$

For technical details the reader is referred to the literature.

(8.17) Theorem (for the fixed boundary problem)
Let $\underline{y}_0$ be an admissible vector function and let $\tilde{\underline{y}}$ be a central field in a field G which is open, connected, and simply connected. Suppose also that $\underline{y}_0$ is embedded in the central field $\tilde{\underline{y}}$; i.e. there is an n-tuple of parameters $\alpha_0 = (\alpha_{01}, \alpha_{02}, \ldots, \alpha_{0n})$ such that for all $x \in [a, b]$, (x, α_0) lies in the domain of definition M of $\tilde{\underline{y}}$ and $\underline{y}_0(x) = \tilde{\underline{y}}(x, \alpha_0)$. Furthermore, suppose that for all $(q_i) \in \mathbb{R}^n$ and all $(x, y_i) \in G$,

$$\mathcal{E}(x, y_i,\ p_i(x, y_j),\ q_i) \geq 0.$$

Then for all admissible functions $\underline{y}$ whose graph lies in the field G, $J(y) \geq J(y_0)$. In particular, $\underline{y}_0$ is a strong solution of the variational problem. This theorem is the analogue of Theorem (6.11).

(8.18) Theorem (for fixed boundary problems)
Let y_0 be an extremal which has smooth fourth derivatives and which satisfies the boundary conditions of the fixed boundary problem. Let $\underline{y}_0$ be regular and suppose the interval $[a, b]$ contains no points conjugate to a. Then if $\underline{y}_0$ satisfies Legendre's necessary condition (8.11), *$\underline{y}_0$ is a weak solution of the variational problem.*

This is analogous to the supplement (6.18) to Theorem (6.12).

Example

We continue the investigation of the example from Section 8.1, where the variation integral is

$$J(y) = \int_0^1 [y_1'^2(x) + (y_2'(x) - 1)^2 + y_1^2(x) + y_1(x)y_2(x)]\, dx.$$

It can be readily shown that the integrand is not a convex function of y_1, y_2, y_1', y_2'. The Euler excess function $\mathcal{E}$ is expressed by

$$\begin{aligned}\mathcal{E}(x, y_1, y_2, y_1', y_2', q_1, q_2) &= q_1^2 + (q_2 - 1)^2 - y_1' - (y_2' - 1)^2 \\ &\quad -2y_1'(q_1 - y_1') - 2(y_2' - 1)(q_2 - y_2') \\ &= (q - y_1')^2 + [q_2 - 1 - (y_2' - 1)]^2 \geq 0.\end{aligned}$$

If we take the boundary conditions to be

$$\underline{y}(0) = 0, \quad \underline{y}(1) = 0, \tag{8.19}$$

then the null vector $\underline{y}_0 = 0$ is the only extremal which satisfies (8.19). With the help of Theorem (8.17) we can show that $\underline{y}_0 = 0$ is also a solution of the variational problem with the boundary conditions (8.19). To this end, we need a central field $\tilde{\underline{y}}$ in which $\underline{y}_0$ (for $x \in [0,1]$) is embedded. The extremals $y(\cdot, A, C,) = (y_1(\cdot, A, C),\ y_2(\cdot, A, C))$, defined by

$$\begin{aligned}y_1(x, A, C) &= 2Ak_1^2 \sinh(k_1 x + B_0) - 2Cm^2 \sin(mx + D_0),\\ y_2(x, A, C) &= A \sinh(k_1 x + B_0) + C \sin(mx + D_0),\\ &\quad (A \text{ and } C \text{ are family parameters})\end{aligned}$$

pass through the centre $(x_0, 0)$; i.e. $y(x_0, A, C) = 0$ for all A and C if

$$k_1 x_0 + B_0 = 0 = m x_0 + D_0.$$

Clearly $\underline{y}_0 = 0 = y(\cdot, 0, 0)$ is an extremal of the family. For our example, we choose $x_0 < 0$ and so that $|x_0|$ is sufficiently small and $B_0 = -k_1 x_0$, $D_0 = -m x_0$. It is also possible to take $x_0 = 0$, but then one must employ a limiting process. It remains to check whether $\underline{y}(\cdot, A, C)$ has properties (ii) and (iii) of a control field (cf. Definition (8.15)).

Consider (ii) first:

$$\det \frac{\partial(y_1, y_2)}{\partial(A, C)} = \det \begin{pmatrix} 2k_1^2 \sinh k_1(x - x_0) & 2m^2 \sin m(x - x_0) \\ \sinh k_1(x - x_0) & \sin m(x - x_0) \end{pmatrix}$$
$$= 2(k_1^2 - m^2) \sin m(x - x_0) \sinh k_1(x - x_0) \neq 0,$$
$$\left(k_1 = \sqrt{\tfrac{1}{2} + \sqrt{\tfrac{1}{2}}},\ m = \sqrt{\sqrt{\tfrac{1}{2}} - \tfrac{1}{2}} \right), \tag{8.20}$$

provided $k_1(x - x_0) \neq 0$ and $m(x - x_0)$ is not an integral multiple of π. Since the domain of definition of the admissible functions is $[0, 1]$, we must have $0 < m(x - x_0) < \pi$ for all x such that $x_0 < x \leq 1$. With a proper choice of x_0 this is easily satisfied.

Condition (iii): Here we must show that there is a set G in the xy_1y_2-space such that through each point $(\tilde{x}, \tilde{y}_1, \tilde{y}_2)$ of G one and only one extremal of the family $y(\cdot, A, C)$ passes. To do this, set

$$(\tilde{x}, \tilde{y}_1, \tilde{y}_2) = (\tilde{x}, \underline{y}(\tilde{x}, A, C)) \tag{8.21}$$

to determine the parameters A and C so that (8.21) is satisfied:

$$\tilde{y}_1 = y_1(\tilde{x}, A, C), \quad \tilde{y}_2 = y_2(\tilde{x}, A, C).$$

This is a system of linear equations for the desired parameters A and C. The determinant of the coefficient matrix of this system of equations is (8.20); it is not equal to zero for $x \in [0, 1]$. Thus the system is solvable. We can therefore conclude that the extremals of the family $\underline{y}(\cdot, A, C)$ form a central field with the domain of definition

$$M = \{(x, A, C) : x_0 < x < x_0 + \frac{\pi}{m},\ A, C \text{ arbitrary}\}$$

and with the field

$$G = \{(x, y_1, y_2) \in \mathbb{R}^3 : x_0 < x < x_0 + \frac{\pi}{m},\ y_1, y_2 \in \mathbb{R}\}.$$

We have shown, therefore, that the assumptions of Theorem (8.17) are satisfied, and we may conclude that there is a central field in which $\underline{y}_0$ is embedded and that the Weierstrass excess function $\mathcal{E}$ is nowhere negative. It follows that $\underline{y}_0$ is a solution of the variational problem. That is, $J(y) \geq J(y_0)$ holds for all admissible functions $\underline{y}$ for the fixed boundary problem stated above.

9

The parametric problem

In the four examples of the first chapters we sought to find curves or paths with minimum or maximum properties. In order to obtain a simple mathematical formulation of the problem, we restricted ourselves in Chapter 1 to those curves which are graphs of a function y. In order to solve the original problem without such restriction, certain additional considerations (for the most part difficult) are needed. Furthermore, the procedures are often impractical and sometimes in principle impossible. We want now, however, to define the variational problem for a broader class of curves and thereby define the so-called *parametric problem.* Subsequently, we will derive necessary and sufficient conditions for its solution.

9.1 STATEMENT OF THE PROBLEM

In this section we want to clarify the concept of the parametric representation of a curve and of the parametric transformation, and show how the variational integral for the curve must look. Finally, we will define the distance $d_0(k_1, k_2)$ and $d_1(k_1, k_2)$ for two curves k_1 and k_2.

(9.1) Definition

An *oriented m-fold smooth or piecewise smooth curve* k in $\mathbb{R}^n$ $(n = 2, 3, \ldots)$ is defined to be an m-fold smooth or piecewise smooth vector function $\underline{z}$, respectively, with the property that $\underline{z}'(t) \neq 0$ for all t of the domain of definition $[t_a, t_b]$ of $\underline{z}$. The vector function $\underline{z}$ is then called the *C^m-parametric representation* or the *D^1-parametric representation*, respectively, of the curve k, and t is called the *parameter.*

Suppose now that $\underline{z} : [t_a, t_b] \mapsto \mathbb{R}^n$ and $\underline{z}^* : [u_a, u_b] \mapsto \mathbb{R}^n$ are two m-fold smooth vector functions with $z'(t) \neq 0$ and $z^{*\prime}(u) \neq 0$ for all t and u. Then $\underline{z}$ and $\underline{z}^*$ are C^m-parametric representations of the same oriented curve k if there is an invertible transformation $g : [t_a, t_b] \mapsto [u_a, u_b]$ of the parameter which has the following properties:

(i) g and its inverse function $h : [u_a, u_b] \mapsto [t_a, t_b]$ are m-fold smooth functions;

(ii) $g'(t) > 0 \quad$ for all $\quad t \in [t_a, t_b]$;

(iii) $\underline{z}(t) = \underline{z}^*(g(t)) \quad$ for all $\quad t \in [t_a, t_b] \quad$ and $\quad \underline{z}^*(u) = \underline{z}(h(u))$ for all $\quad u \in [u_a, u_b]$.

We call g the *C^m-parametric transformation*. The parametric transformation for piecewise smooth curves can be defined analogously.

All parametric representations $\underline{z}^*$ obtained from a given parametric representation by means of a parametric transformation g are, from the theoretical standpoint, of the same value. From the practical point of view, one representation may be much more satisfactory than all the others; for one special parametric representation of the curve being considered, the calculations may be much more convenient. We shall regard the values of the function $\underline{z}$ to be properties of the curve which are independent of the special choice of parametric representation, and such that each remains unchanged for all choices of the parameter. Thus, for example, the path $\{\underline{z}(t) : t \in [t_a, t_b]\}$ of the curve with the parametric representation $\underline{z}$, the initial point A and the endpoint B of the curve, the tangent to the curve, etc., are independent of the parametric representation. A variational integral for curves must be defined with the help of the parametric representation, but its values, which are properties of the curve, will be independent of the special choice of parameter.

Every m-fold smooth vector function $\underline{y} : [a, b] \mapsto \mathbb{R}^n$ defines a curve in $\mathbb{R}^{n+1}$ for which x is the parameter. The transformation

$$x \in [a, b] \mapsto (x, \underline{y}(x)) \in \mathbb{R}^{n+1} \tag{9.2}$$

is a C^m-parametric representation, and the graph of the vector function $\underline{y}$ is the path.

Curves with a parametric representation of the kind (9.2) have the property that the projection of the path on the x-axis is an invertible, single-valued transformation. In general, projections on any one of the axes are not invertible, single-valued functions. Consider, for example, the circle. But it can be shown that a sufficiently short piece of a curve will always have a parametric representation of the type (9.2). Usually one has only to rename the axes.

In order to motivate the definition of the variational integral $J(k)$ for a curve k, we suppose first that k is an oriented curve which has a D^1-parametric representation of the type (9.2): $(x, \underline{y}(x))$. The variational integral for the function $\underline{y}$ has the usual form

$$J(y) = \int_a^b f(x, \underline{y}(x), \underline{y}'(x))\, dx.$$

Let $\underline{z} = (z_1, z_2, \cdots, z_{n+1}) : [t_a, t_b] \mapsto \mathbb{R}^{n+1}$ be a D^1-parametric representation of the oriented curve k. Then the parametric transformation which carries the parameter t to the parameter x will describe the curve through $g = z_1$ so that

$$x = z_1(t),\ y_1(x) = z_2(t), \ldots, y_n(x) = z_{n+1}(t),\ g'(t) = z_1'(t) > 0.$$

Because $z_i'(t) = y_{i-1}'(g(t))g'(t)$, the change of parameter yields for the integral $J(\underline{y})$ the expression

$$J(\underline{y}) = \int_{t_a}^{t_b} f\left(z_1(t), z_i(t), \frac{z_i'(t)}{z_1'(t)}\right) z_1'(t)\, dt \quad (i = 2, \ldots, n+1)$$

$$= \int_{t_a}^{t_b} F(z_j(t), z_j'(t))\, dt \quad (j = 1, \ldots, n+1).$$

Here we have set

$$F(z_j, z_j') = f\left(z_1, z_i, \frac{z_i'}{z_1'}\right) z_1'; \tag{9.3}$$

$$i = 2, 3, \ldots, n+1; \quad j = 1, 2, \ldots, n+1; \quad z_1' > 0,$$

where z_j and z_j' denote variables. The integrand F is independent of the curve parameter t and has the property

$$F(z_j, cz_j') = cF(z_j, z_j') \quad \text{if} \quad c > 0. \tag{9.4}$$

Thus F is a *positive homogeneous function of the first order* in the variable z_j' and

$$\begin{aligned} F(z_j, cz_j') &= f\left(z_1, z_i, \frac{cz_i'}{cz_1'}\right) cz_1' \\ &= cf\left(z_1, z_i, \frac{z_i'}{z_i}\right) z_1' \\ &= cF(z_j, z_j'). \end{aligned}$$

The converse can be shown easily.

(9.5) Theorem

For G open in $\mathbb{R}^n$, let $F : G \times \mathbb{R}^n \mapsto \mathbb{R}$ be a given continuous function which is positive homogeneous of first order in the last n variables z_j. Suppose also that the oriented curve k has the D^1-parametric representations $\underline{z} : [t_a, t_b] \mapsto \mathbb{R}^n$ and $\underline{z}^* : [u_a, u_b] \mapsto \mathbb{R}^n$. Then

$$\int_{t_a}^{t_b} F(\underline{z}(t), \underline{z}'(t))\, dt = \int_{u_a}^{u_b} F(\underline{z}^*(u), \underline{z}^{*'}(u))\, du. \tag{9.6}$$

The integral (9.6) thus depends on the curve k and so will be denoted by $J(k)$. The assumptions on the integrand F of the variational integral $J(k)$ of a parametric problem are the following:

(i) F is defined and continuous on $G \times \mathbb{R}^n$, where G is an open set in R^n;

(ii) F is positive homogeneous of the first order in the last n variables z_j'; and

(iii) F is three-fold smooth at all points (z_j, z_j') in its domain of definition, provided not all $z_j' = 0$.

Admissible functions are all oriented, piecewise smooth curves k whose paths lie in G and which satisfy certain boundary conditions. (Initial and endpoints are given for the fixed boundary problem; initial point is given and the endpoint lies on a prescribed boundary curve for the problem between point and curve, etc.) The distances d_0 and d_1 can also be defined for these curves:

$$\begin{aligned} d_0(k_1, k_2) = \inf\{d_0(\underline{z}_1, \underline{z}_2) : \; & \underline{z}_1 \text{ and } \underline{z}_2 \text{ are, respectively, the} \\ & D^1\text{-parametric representations of } k_1 \text{ and } k_2 \\ & \text{with a common domain of definition } [t_a, t_b]\} \;, \\ d_1(k_1, k_2) = \inf\{d_1(\underline{z}_1, \underline{z}_2) : \; & \underline{z}_1 \text{ and } \underline{z}_2 \text{ as for } d_0 \;\}. \end{aligned}$$

With the above definitions the concepts of 'solution', 'weak solution' and 'strong solution' of a parametric problem can be defined as they were previously.

9.2 NECESSARY AND SUFFICIENT CONDITIONS

The parametric problem consists of finding, among all the piecewise smooth parametric representation $\underline{z}$ of curves k which satisfy the boundary conditions, those for which the integral

$$J(k) = \int_{t_a}^{t_b} F(\underline{z}(t), \underline{z}'(t))\, dt$$

has a minimal value. But because each admissible curve may be represented parametrically by a vector function $\underline{z}$ with a prescribed interval of definition $[t_a, t_b]$ (or $[0,1]$), the curve problem $J(k) \to \min$ (with boundary conditions) is reduced to a variational problem where the function $\underline{z} : [t_a, t_b] \mapsto \mathbb{R}^n$ is to be found. It follows from this that necessary conditions like those of Chapters 8 and 11 are valid; i.e. the Euler differential equation, the corner conditions, the Legendre and Weierstrass necessary condition (where we write t instead of x, z_i instead of y_i and where F replaces f) must all be satisfied. The above reasoning, however, breaks down with respect to Jacobi's necessary condition. For the parametric problem Jacobi's necessary condition has the same meaning as it has for the variational problem with one or several functions to be found: only sufficiently short 'pieces' of extremals can be solutions. We will forego here, therefore, the development of Jacobi's necessary condition for reasons which will be put forth in the next section. The sufficient conditions for the fixed boundary problem with convex integrand may also be carried over directly (cf. Theorem 8.9).

As an example, consider the problem of the shortest curve joining points A and B on a surface F (cf. Fig. 9.1). We assume the surface has a four-fold smooth (partially) parametric representation

$$\underline{x}(u_1, u_2) = (x_1(u_1, u_2), x_2(u_1, u_2), x_3(u_1, u_2)).$$

A piecewise smooth curve on the surface can be represented, then, as a function of the parameter t:

$$\underline{z}(t) = (z_1(t), z_2(t), z_3(t)) = \underline{x}(u_1(t), u_2(t)),$$

where z_i, $i = 1, 2, 3$, and u_1 and u_2 are piecewise smooth functions. The length L of the curve can be determined from the formula

$$L = \int_{t_a}^{t_b} |z'(t)| \, dt.$$

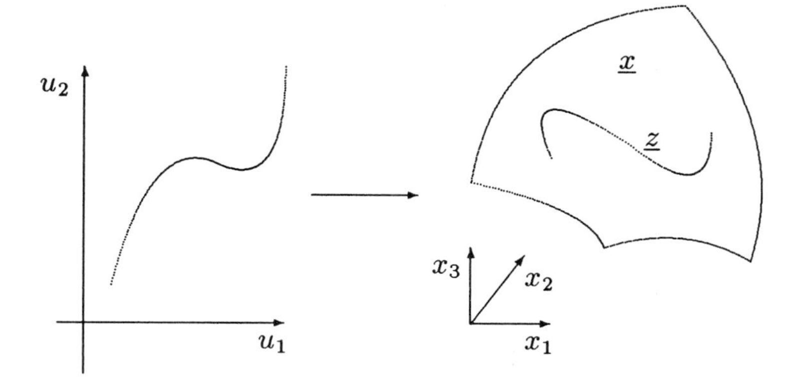

Fig. 9.1. Parametric representation of a surface.

But the square of the length of the vector $\underline{z}'(t)$ is

$$\begin{aligned}|\underline{z}'(t)|^2 &= \underline{z}'(t)\cdot\underline{z}'(t)\\ &= \sum_{i=1}^{2}\underline{x}_{u_i}u_i'(t)\cdot\sum_{j=1}^{2}\underline{x}_{u_j}u_j'(t)\\ &= \sum_{i,j=1}^{2}\underline{x}_{u_i}\cdot\underline{x}_{u_j}u_i'(t)u_j'(t).\end{aligned}$$

If we let g_{ij} denote the scalar product of the vectors $\underline{x}_{u_i}$ and $\underline{x}_{u_j}$, then L is represented by

$$L=\int_{t_a}^{t_b}\left[\sum_{i,j=1}^{2}g_{ij}(u_1(t),u_2(t))u_i'(t)u_j'(t)\right]^{1/2}dt.$$

The function

$$F(u_1,u_2,u_1',u_2')=\left[\sum_{i,j=1}^{2}g_{ij}(u_1,u_2)u_i'u_j'\right]^{1/2} \tag{9.7}$$

satisfies the assumptions on the integrand of the parametric problem so that

$$F_{u_i'}=\frac{1}{F}\sum_{j=1}^{2}g_{ji}u_j',\quad F_{u_i}=\frac{1}{2F}\sum_{j,k=1}^{2}\frac{\partial g_{jk}}{\partial u_i}u_j'u_k'.$$

(We have omitted the argument here.) The parameter t represents arc length if $F(u_i(t), u_i'(t)) = 1$ for all t. If t is such a parameter, then the Euler differential equation has the following simple form:

$$\sum_{j=1}^{2} g_{ij}(\underline{u}(t))u_j''(t) + \sum_{j,k=1}^{2} \left[\frac{\partial g_{ij}}{\partial u_k}(\underline{u}(t)) - \frac{1}{2}\frac{\partial g_{jk}}{\partial u_i}(\underline{u}(t))\right] u_j'(t)u_k'(t) = 0 \tag{9.8}$$

$$(i = 1, 2).$$

Legendre's necessary condition appears as

$$\left[\sum_{i,k=1}^{2} g_{ik}(\underline{u}(t))u_i'(t)v_k\right]^2 + \sum_{i,k=1}^{2} g_{ik}(\underline{u}(t))v_iv_k$$
$$\geq 0 \text{ for all } (v_1, v_2) \in \mathbb{R}^2.$$

Since the second term is equal to or greater than zero, Legendre's necessary condition is always satisfied. The integrand F is generally not a convex function of the arguments u_1, u_2, u_1', u_2'. So much for this example.

Exercises

(9.9) Find the shortest curve joining prescribed initial and endpoints A and B on the slit circular cylinder whose parametric representation is

$$\underline{x}(u_1, u_2) = (r\cos u_1, r\sin u_1, u_2),$$

$$0 < u_1 < 2\pi, \;\; u_2 \in \mathbb{R}.$$

Hint: The integrand $F(u_1, u_2, u_1', u_2') = \sqrt{r^2 {u_1'}^2 {u_2'}^2}$. It needs to be shown that the helices, which can be represented by

$$\underline{z}(t) = (r\cos(c_1t + c_2), \;\; r\sin(c_1t + c_2), h_1t + h_2),$$

where c_1, c_2, h_1 and h_2 are constants and $r^2c_1^2 + h_1^2 \neq 0$, together with the special cases of lines and circles, are the extremals of the parametric problem. What changes arise if the slit cylinder is replaced by the complete circular cylinder? How many extremals are there then for the given boundary points?

(9.10) Formulate Example (1.2) (surface of revolution) and Example (1.3) (Bernoulli) as parametric problems without the restriction that x be chosen as the curve parameter. Find the Euler equation of the parametric problem. It is convenient to choose the arc length σ as the parameter. With this choice

$$\left(\frac{dx}{d\sigma}\right)^2 + \left(\frac{dy}{d\sigma}\right)^2 = 1,$$

a condition which essentially determines σ uniquely. This Euler differential equation can be solved (cf. Theorem (9.15)).

9.3 SOME PARTICULARS OF THE PARAMETRIC PROBLEM AND PROOF OF THE SECOND ERDMANN CORNER CONDITION

The impression can easily arise from the study of Section 9.2 that with respect to applications a parametric problem is only an interesting special case of a variational problem with n unknown functions.

As we will see in this section, this impression is misleading. This comes about because, for the derivation of the sufficient condition with the help of fields of extremals as well as for Jacobi's condition, the assumption was essentially that the extremals under investigation be regular. For the parametric problem, however, there are in principle no regular extremals. We must, therefore, develop a new idea.

We will not discuss here the Jacobi condition nor the sufficiency condition. We will only derive a few important formulas for use in connection with the parametric problem. From these it will follow that the Euler differential equations are not independent from one another. Finally, we will derive the second Erdmann corner condition.

We say that a function $F : \mathbb{R}^m \mapsto \mathbb{R}$ is positive homogeneous of the rth order, $r = 0, +1, -1, +2, -2, +3, -3, \ldots$, if

$$F(cx_j) = c^r F(x_j) \quad \text{if} \quad c > 0, \quad j = 1, 2, \ldots, m. \tag{9.11}$$

Differentiation of equation (9.11) with respect to x_r, c held constant, $c > 0$, yields

$$cF_{x_k}(cx_j) = c^r F_{x_k}(x_j),$$

which implies that the partial derivatives of F with respect to x_k are positive-homogeneous of order $(r-1)$.

If, on the other hand, (9.11) is differentiated with respect to c (x_j held fixed) at the point $c = 1$, then we obtain

$$\sum_{k=1}^{n} F_{x_k}(x_j) x^k = rF(x_j), \tag{9.12}$$

the so-called *Euler homogeneity relation.* The integrand F of a parametric problem is positive homogeneous of first order in the last n variables z_i. Thus F'_{z_k} is positive homogeneous of the zeroth order so that the following homogeneity relations hold:

$$\sum_{k=1}^{n} F_{z'_k}(z_j, z'_j) z'_k = F(z_j, z'_j), \tag{9.13}$$

$$\sum_{k=1}^{n} F_{z'_i z'_k}(z_j, z'_j) z'_k = 0 \quad \text{for} \quad i = 1, \ldots, n. \tag{9.14}$$

From (9.14) it follows that (z'_i) is an eigenvector for the eigenvalue 0 of the matrix $(F_{z'_i z'_k}(z_j, z'_j))_{i,\,k=1,\cdots,n}$. Thus the determinant of this matrix is 0. In Section 8.2 we said that a vector function $\underline{z}$ is called regular if the determinant of the matrix $F_{z'_i z'_k}(\underline{z}(t), \underline{z}'(t))$ is different from 0 for all t. Because F is positive homogeneous, this condition of regularity in the case of the parametric problem is always vulnerable. Such is the price paid for the variety of parametric representations of curves.

With the help of the homogeneity conditions (9.14) and since the partial derivatives F_{z_i} of F are positive homogeneous of first order, we can prove the following theorem.

(9.15) Theorem

The n Euler differential equations for the parametric problem are not independent of each other. Indeed, if $n-1$ of the equations are satisfied, then the last is also satisfied.

This theorem is closely related to the fact that if there is a solution $\underline{z}$ for the Euler differential equation, then it is never uniquely determined, because all other smooth parametric representations of the curve represented by $\underline{z}$ are also solutions of the Euler equation. Because of Theorem (9.15) one can use the following technique to solve the system of Euler equations. Replace one of the equations (8.5) (the most complex or intractable) with a broad, arbitrary condition by means of which the curve parameter is determined. So, in the example of Section 9.2, in which the shortest curve joining two points on a surface is sought, equation (9.8) can be replaced by the condition $F(\underline{u}(t), \underline{u}'(t)) = 1$.

Proof of Theorem (9.15)

If the implied differentiation is carried out, the Euler differential equations have the form

$$F_{z_k} - \sum_{i=1}^{n} [F_{z'_k z_i} z'_i(t) + F_{z'_k z'_i} z''_i(t)] = 0. \tag{9.16}$$

(The arguments for the partial derivatives of F are $\underline{z}(t)$ and $\underline{z}'(t)$.)

Since for all t it holds that $\underline{z}'(t) \neq 0$ (cf. definition of the C^1-parametric representation), at least one of the components of $\underline{z}'(t)$, let us say the nth component $z'_n(t)$, is different from zero.

We multiply the relations (9.14)

$$\sum_{k=1}^{n-1} F_{z_i' z_k'}(\underline{z}(t), \underline{z}'(t)) z_k'(t) = -z_n'(t) F_{z_i' z_n'}(\underline{z}(t), \underline{z}'(t)),\ i = 1, \cdots, n$$

with $z_i''(t)$, respectively, and then add the resulting n equations to obtain

$$-z_n'(t) \sum_{i=1}^{n} F_{z_i' z_n'} z_i''(t) = \sum_{i=1}^{n} \sum_{k=1}^{n-1} F_{z_i' z_k'} z_k'(t) z_i''(t). \qquad (9.17)$$

Now assume that $\underline{z}$ satisfies the first $n-1$ Euler differential equations (9.16). Then the right-hand side $\sum_{i,k} F_{z_i' z_k'} z_k' z_i''$ of (9.17) can be replaced by

$$\sum_{k=1}^{n-1} \left[F_{z_k} - \sum_{i=1}^{n} F_{z_k' z_i} z_i'(t) \right] z_k'(t).$$

Because F_{z_k} is positive homogeneous of first order, the right-hand side of (9.17) is equal to

$$\begin{aligned} &\sum_{k=1}^{n-1} \sum_{i=1}^{n} [F_{z_k z_i'} z_i'(t) - F_{z_k' z_i} z_i'(t)] z_k'(t) \\ &\qquad = \sum_{i,k=1}^{n} [F_{z_k z_i'} - F_{z_k' z_i}] z_i'(t) z_k'(t) \\ &\qquad\qquad - \sum_{i=1}^{n} [F_{z_n z_i'} - F_{z_k' z_i}] z_n'(t) z_i'(t) \\ &\qquad = -z_n'(t) \left[F_{z_n} - \sum_{i=1}^{n} F_{z_n' z_i} z_i'(t) \right]. \end{aligned}$$

In summary we can write

$$-z_n'(t) \sum_{i=1}^{n} F_{z_i' z_n'} z_i''(t) = -z_n'(t) \left[F_{z_n} - \sum_{i=1}^{n} F_{z_n' z_i} z_i'(t) \right].$$

The arguments for each partial differentiation are $\underline{z}(t)$ and $\underline{z}'(t)$.

Since $z_n'(t) \neq 0$, the last equation may be divided by $-z_n'(t)$. We have thus shown that a smooth vector function $\underline{z}$, whose

nth component is not zero at $t = 0$ and which satisfies the first $n-1$ Euler differential equations, must also satisfy the nth Euler differential equation.

We turn next to the proof of the *second Erdmann corner condition* (2.29). In doing so we move from the variational problem with a single unknown function to the parametric problem with the integrand (9.3).

(9.18) Theorem
Suppose the weak solution y^* of the variational problem

$$J(y) = \int_a^b f(x, \underline{y}(x), \underline{y}'(x))\, dx \to \min$$

has a corner at the point $x = x_0$. Then the left- and right-hand derivatives $y_l^*(x_0)$ and $y_r^*(x_0)$, respectively, of y^* satisfy the following conditions:

$$f(x_0, \underline{y}^*(x_0), \underline{y}_l^{*'}(x_0)) - \sum_{i=1}^{n} f_{y_i}(x_0, \underline{y}^*(x_0), \underline{y}_l^{*'}(x_0)) y_{il}^{*'}(x_0)$$

$$= f(x_0, \underline{y}^*(x_0), \underline{y}_r^{*'}(x_0)) - \sum_{i=1}^{n} f_{y_i}(x_0, \underline{y}^*(x_0), \underline{y}_r^{*'}(x_0)) y_{ir}^{*'}(x_0). \tag{9.19}$$

Proof
Since y^* is a weak solution of the variational problem $J(y) \to \min$, it follows that the curve k^* with the parametric representation

$$\underline{z}^*(x) = (z_1^*(x), \ldots, z_{n+1}^*(x)) = (x, y_1^*(x), \ldots, y_n^*(x)),$$

is a weak solution of the parametric problem

$$J(k) = \int_{t_a}^{t_b} F(\underline{z}(t), \underline{z}'(t))\, dt \to \min \quad (z_1' > 0),$$

where (cf. (9.3))

$$F(z_i, z_i') = f\left(z_1, z_2, \ldots, z_{n+1}, \frac{z_2'}{z_1'}, \ldots, \frac{z_{n+1}'}{z_1'}\right) z_1'. \tag{9.20}$$

Just as all curves k which, with respect to the distance function d_1, lie sufficiently near k^* may be represented by the vector function $\underline{y}$, so also may k^* be represented by $\underline{y}^*$ so that $J(k) = J(y)$. Since y^* has a corner at the point x_0, so also has $\underline{z}^*$ a corner there. Thus $\underline{z}^*$ satisfies the corner conditions. The first of these conditions is expressed by

$$F_{z_1'}(\underline{z}^*(x_0), \underline{z}_l^{*'}(x_0)) = F_{z_1'}(\underline{z}^*(x_0), \underline{z}_r^{*'}(x_0)). \tag{9.21}$$

From the homogeneity relation (9.13) it follows that

$$z_1' F_{z_1'}(z_i, z_i') = F(z_i, z_i') - \sum_{j=2}^{n+1} F_{z_j'}(z_i, z_i') z_j',$$

which allows us to replace $F_{z_1'}$ in (9.21) and thus return to the notation of the variational problem $J(y) \to \min$. To do this we need (9.20) and the relationship $z_1^{*'}(x) = 1$ (for all x), which follows from it.

Then we can write

$$F_{z_j'}(z_1, \ldots, z_{n+1}, 1, z_2', \ldots, z_{n+1}') = f_{y_{j-1}'}(z_1, \ldots, z_{n+1}, z_2', \ldots z_{n+1}')$$
$$\text{for} \quad j = 2, \ldots, n+1.$$

This concludes the proof.

We consider next a simple and graphic example for which corners appear.

(9.22) Example

Consider a parametric problem with fixed boundary and for which the integrand does not depend on the variables $z_1, \cdots, z_n$. We will thus write $F(z_i')$ instead of $F(z_i, z_i')$. The extremals are now straight lines. If F is convex, then the line segment joining the endpoints is the solution of the variational problem. For the case where F is not convex, it can be shown that for certain endpoints only solutions with corners can be found. The proof employs the following ideas:

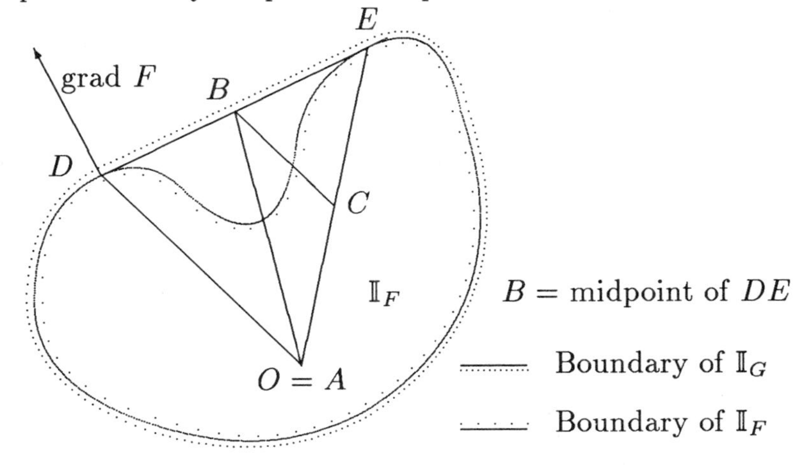

Fig. 9.2. Smallest convex set containing another.

(i) A positive integrand F of a parametric problem is convex if and only if the set of points $\mathbb{I}_F = \{(z_i') \in \mathbb{R}^n : F(z_i') \leq 1\}$, called the *indicatrix*, is convex. (To show this one can use (ii).)

(ii) The graph of a positive homogeneous function F of first order is a generalized cone with its vertex at 0.

(iii) Let G be a positive integrand of a parametric problem which is also independent of z_i. If for all (z_i') it holds that $F(z_i') \geq G(z_i')$, then $\mathbb{I}_F \subset \mathbb{I}_G$; conversely, if $\mathbb{I}_F \subset \mathbb{I}_G$, then $F(z_i') \leq G(z_i')$ for all $(z_i') \in \mathbb{R}^n$.

(iv) The set $\mathbb{I}_F$ is identical with the set of points $B \in \mathbb{R}^n$ which has the property that the variational integral over the path from $A = 0$ to B is equal to or less than 1.

Suppose that $\mathbb{I}_F$ and $\mathbb{I}_G$ are given, as shown in Fig. 9.2. $\mathbb{I}_G$ is the smallest convex set which contains $\mathbb{I}_F$.

According to (i), F is not convex while G is convex. According to (iii), $G \leq F$. The variational integral $J_F(\underline{z})$ with the integrand F for $\underline{z}$ is either greater than the variational integral $J_G(\underline{z})$ or else it holds that $J_F(\underline{z}) = J_G(\underline{z})$. Since G is convex, the line segment $\underline{z}_0$ from $A = 0$ to B (cf. Fig. 9.2) is a solution of the variational problem $J_G(\underline{z}) \to \min$ with the boundary conditions $\underline{z}(t_a) = A$ and $\underline{z}(t_b) = B$, and $J_G(\underline{z}_0) = 1$. Consequently,

$J_F(\underline{z}) \geq J_G(\underline{z}) \geq 1$ for all piecewise smooth vector functions $\underline{z}$ with initial point A and terminal point B. We now introduce a path $\underline{z}^*$ with a corner C and whose variational integral $J_F(\underline{z}^*) = 1$; $\underline{z}^*$ is then a solution of the variational problem.

For the sake of simplicity, let B be the midpoint of the segment DE (cf. Fig. 9.2).

$$\underline{z}^* : [0,1] \mapsto \mathbb{R}^2$$

$$\underline{z}^*(t) = \begin{cases} A + t(E-A) & \text{if } t \in [0, \frac{1}{2}], \\ C + (t - \frac{1}{2})(D-A) & \text{if } t \in [\frac{1}{2}, 1], C = A + \frac{1}{2}(E-A), \end{cases}$$

$$J_F(\underline{z}^*) = \int_0^{1/2} F(E-A)\,dt + \int_{1/2}^1 F(D-A)\,dt = \tfrac{1}{2} + \tfrac{1}{2} = 1.$$

(We have not differentiated here between points and their position vectors.) The direction $E - A$ and $D - A$ of z^* must, of course, be chosen so that the corner condition at C is satisfied: $F_{z_i'}(E-A) = F_{z_i'}(D-A)$. The vector with the components $F_{z_i'}(E-A)$ is grad $F(E-A)$, which is perpendicular to the line tangent to the level curve $F(\underline{z}') = 1$ at the point $E - A$ (cf. Fig. 9.2).

Exercise

Prove the statements (i), (ii), (iii) and (iv).

10

Variational problems with multiple integrals

If a loop of wire (which represents a closed curve) is dipped into a soap solution, a soap film will be formed. The film will represent an area which has the wire loop k as its boundary. This soap film should not be confused with a bubble which is a closed surface without boundary and so subject to different rules than a soap film. Physics demonstrates that the soap film has the form of that surface which, in comparison with all the surfaces which span the wire loop, has the smallest area. The phenomenon of the soap film leads us to the so-called problem of the *minimal surface*. Let k be a closed curve in $\mathbb{R}^3$. Among all surfaces $\mathbb{F}$ which have k as a boundary curve, find those whose surface area $J(\mathbb{F})$ is smallest. In contrast to the variational integral considered thus far, $J(\mathbb{F})$ is a double integral. The unknown function (or functions) is dependent on two variables, is defined on an appropriate two-dimensional set D and must satisfy a boundary condition on the boundary ∂D of D. These are the most important aspects of the problem which we want to treat in these paragraphs.

10.1 THE STATEMENT OF THE PROBLEM AND EXAMPLES

In this section we will formulate the problem involving multiple integrals for one or several unknown functions. We will give the problem of the minimal surface its mathematical form and, as a further example, we will formulate the Hamiltonian principle for a vibrating string.

The variational integral for a surface was defined in the introductory example for the soap film. A surface may be expressed by different parametric representations similar to those used in the case of curves. And just as we did earlier, it is also best here to differentiate between variational problems for surfaces and variational problems for functions, and to treat variational problems for functions next. In so doing, we want first of all to agree which functions are admissible for consideration. We will assume that the common domain of definition $\overline{D}$ for all admissible functions is the closure of a region D in $\mathbb{R}^m$ $(m = 1, 2, \ldots)$. Furthermore, the boundary shall be assumed smooth.

Definition

A non-empty subset D of $\mathbb{R}^m$ is called a *region* if it is open and connected. We say that a region $D \subset \mathbb{R}^2$ has a *smooth boundary* ∂D if the boundary ∂D is a continuously differentiable curve (more precisely, the path of one such curve) which has a tangent at each point. The curve ∂D is not permitted to have cusps. Correspondingly, we say that a region D in $\mathbb{R}^m$, $m > 2$, has a smooth boundary if the boundary ∂D of D is an $(m-1)$ -dimensional surface which has at each point an $(m-1)$-dimensional tangent plane. The boundary of a region $D \subset \mathbb{R}^m$ with smooth boundary has, therefore, no corners, edges, etc. By the *closure* $\overline{D}$ of a region D we mean the set $\overline{D} = D \bigcup \partial D$. The points of $\overline{D}$ will be denoted by $(x_1, x_2, \cdots, x_m)$ or, more concisely, (x_α).

We shall understand admissible functions to be all functions y

(or vector functions $\underline{y} : \overline{D} \mapsto \mathbb{R}^n$, $\underline{y} = (y_1, y_2, \ldots, y_n)$) which are defined on $\overline{D}$ and which on $\overline{D}$ are smooth. For the fixed boundary problem, the values of y (or $\underline{y}$) must agree on the boundary ∂D with the values of a prescribed boundary function. There are problems involving several integrals for which the boundary is variable; we will not discuss these problems here. We have imposed relatively strong restrictions on the admissible functions in order to avoid deviating from our proper subject matter because of secondary difficulties. These restrictions are, nonetheless, unpleasant for applications. In this connection, it should be made clear that a rectangle in $\mathbb{R}^2$ is not permitted as a domain of definition $\overline{D}$ because its boundary has corners. For general problems which are beyond the scope of this introduction to the calculus of variations, we refer the reader to the more advanced literature, [10], for example.

For the problem at hand, the variational integral is expressed by

$$J(y) = \int_D f(x_\alpha, y(x_\alpha), y_{x_1}(x_\alpha), \ldots, y_{x_m}(x_\alpha))\, dx_1\, dx_2 \cdots dx_m, \tag{10.1}$$

where y is the function sought, and by

$$J(y_i) = \int_D f(x_\alpha, y_i(x_\alpha),\ y_{ix_\alpha}(x_\beta))\, dx_1 \cdots dx_m, \qquad i = 1, 2, \ldots, n; \quad \alpha, \beta = 1, 2, \ldots, m, \tag{10.2}$$

where n functions $y_1, y_2, \ldots, y_n$ are sought. (We will continue to represent these by a vector function $\underline{y}$.)

The integrand f of the variational problem shall be three-fold smooth. We denote the variables with x_α, y, y_α (or $x_\alpha, y_i, y_{i\alpha}$, respectively).

An admissible function y_0 or an admissible vector function $\underline{y}_0$ is called a solution of this variational problem if, for all admissible functions y (or vector functions $\underline{y}$), it holds that

$$J(y_0) \leq J(y) \quad (\text{or } J(\underline{y}_0) \leq J(\underline{y})).$$

We can also speak of relatively strong and weak solutions. To do so, however, we need the help of distance functions d_0 and d_1 to describe what is meant when we say that one admissible function is in the neighbourhood of another. Thus, for example, for two vector functions $\underline{y}_1$ and $\underline{y}_2$,

$$d_0(\underline{y}_1, \underline{y}_2) = \sup_{(x_\alpha)\in D} |\underline{y}_1(x_\alpha) - y_2(x_\alpha)|.$$

In the derivation of the Euler differential equation in Chapter 2, integration by parts was employed. In the case of multiple variational integrals, the theorem of Gauss (or Stokes) is needed for the corresponding step. We will simply cite the theorem here and refer the reader to the textbooks in analysis for the proof.

(10.3) Gauss Integral Theorem

Let $D \subset \mathbb{R}^m$ be a region with smooth boundary ∂D and let $\underline{u} = (u_\alpha) : \overline{D} \to \mathbb{R}^m$ be a vector function which is smooth on $\overline{D} = D \bigcup \partial D$. Then

$$\int_D \sum_{\alpha=1}^{m} \frac{\partial u_\alpha}{\partial x_\alpha}(x_\beta)\, dx_1 \cdots dx_m = \int_D \operatorname{div} \underline{u}(x_\alpha)\, dx_1 \cdots dx_m$$
$$= \int_{\partial D} \underline{u}(p) \cdot \underline{n}(p)\, d\sigma(p),$$

where div $\underline{u}$ *is the divergence of* $\underline{u}$*,* $d\sigma(p)$ *is the upper surface element of area for* ∂D *at the point* $p \in \partial D$*, and* $\underline{n}(p)$ *is the unit vector which is perpendicular to* ∂D *and is directed outwardly (exterior normal).*

We are now ready to formulate the problem of the minimal surface. Under certain assumptions and restrictions we obtain a variational integral with one unknown function. In the general case we encounter three unknown functions, namely, the component functions of the parametric representation of the surface. For the first and simpler formulation we assume that the orthogonal projection of the prescribed (boundary) curve k on the xy-plane is a smooth curve k^* which is the boundary curve of a region D in the xy-plane. The orthogonal projection shall also determine an invertible, one-to-one correspondence between points of k and k^*.

Consider now functions z of the variables x and y, $z : D \mapsto \mathbb{R}$, which are smooth (partially) with respect to x and y. The graph $\{(x, y, z(x,y)) : (x,y) \in D\}$ is a surface $\mathbb{F}_z$; for the functions z considered here, $\mathbb{F}_z$ shall have the given curve k as a boundary curve. It is clear that not every surface spanning k can be represented as $\mathbb{F}_z$ with the help of a function z (because the orthogonal projection on the xy-plane is not one-to-one for each surface). If we restrict the set of admissible functions with the additional requirement that each be representable by $\mathbb{F}_z$ with the help of a differentiable function, then the variational integral J, which is the area of the surface $\mathbb{F}_z$, is given by

$$J(z) = \int_D \sqrt{1 + z_x^2(x,y) + z_y^2(x,y)}\, dx\, dy. \tag{10.4}$$

In the general case we use a parametric representation $\underline{x}$ of the surface $\mathbb{F}$, where $\underline{x}$ is defined on a unit circle D

$$\underline{x}(u_1, u_2) = (x_1(u_1,u_2),\ x_2(u_1,u_2),\ x_3(u_1,u_2))$$
$$\text{for all } (u_1,u_2) \in \mathbb{R}^2 \text{ with } u_1^2 + u_2^2 \leq 1.$$

We assume that the functions x_i are smooth. Then the formula for the surface area $J(\underline{x})$ of the surface $\mathbb{F}$ represented parametrically is given by

$$J(\underline{x}) = \int \sqrt{g(u_1,u_2)}\, du_1\, du_2,$$

where

$$g(u_1,u_2) = |\underline{x}_{u_1}(u_1,u_2)|^2 |\underline{x}_{u_2}(u_1,u_2)|^2 - (\underline{x}_{u_1}(u_1,u_2) \cdot \underline{x}_{u_2}(u_1,u_2))^2,$$

and $\underline{x}_{u_\alpha}$ is the partial derivative of the vector function $\underline{x}$ with respect to the variables $u_\alpha, \alpha = 1, 2$.

The study of the *vibration of a string* leads also to a variational problem with a double integral. Suppose a homogeneous string with mass ρ per unit length is attached to fixed points at its ends and is held under constant tension μ. We choose the x-axis of a Cartesian coordinate system so that when the string is in its rest position, it lies along the x-axis. The ends are at $A = (0,0)$ and $B = (\ell, 0)$, where ℓ is the length of the string at rest. Let $y = y(x,t)$ be the displacement of the string from its rest position at the point x and at time t. The kinetic energy $T(t)$ and the potential energy $U(t)$ of the string at time t are given by

$$T(t) = \frac{1}{2} \int_0^\ell \rho \left(\frac{\partial y}{\partial t} \right)^2 (x,t)\, dx$$

and

$$\begin{aligned} U(t) &= \mu\ [(\text{length of string at time } t) - (\text{ length at rest})] \\ &= \mu \left[\int_0^\ell \sqrt{1 + y_x^2(x,t)}\, dx - \ell \right] \\ &= \mu \int_0^\ell \left(\sqrt{1 + y_x^2(x,t)} - 1 \right) dx. \end{aligned}$$

Hamilton's principle states that the displacement y behaves in such a way that the action integral is stationary; i.e.

$$\begin{aligned} J(y) &= \int_{t_0}^{t_1} (T(t) - U(t))\, dt \\ &= \int_{t_0}^{t_1} \int_0^\ell \left[\tfrac{1}{2} \rho y_t^2(x,t) + \mu - \mu \sqrt{1 + y_x^2(x,t)} \right] dx\, dt. \end{aligned} \tag{10.5}$$

The domain of definition D of y is the rectangle $\{(x,t) : 0 \leq x \leq \ell,\ t_0 \leq t \leq t_1\}$. How can the boundary conditions be stated? The condition that the ends of the string be held fast means that $y(0,t) = 0,\ y(\ell,t) = 0$ for all t. The function $y(x, t_0)$, which represents the displacement of the string at the moment the oscillation begins, is a given function of x. A boundary condition of the kind $y(x, t_1) = \varphi(x)$, where $\varphi(x)$ is a prescribed function, is not sensible for physical means. This part of the boundary condition may be replaced by the change $y_t(x, t_0)$ of the displacement with respect to time at the initial time t_0 of the oscillation. (By providing $y_t(x, t_0)$, the kinetic energy $T(t_0)$ of the string at time t_0 is determined.)

10.2 THE EULER DIFFERENTIAL EQUATIONS

As usual, we begin with the assumption that y_0 (or $\underline{y}_0$) is a two-fold smooth (weak) solution of the variational problem. By means of a smooth function $Y : D \mapsto \mathbb{R}$ (or vector function $\underline{Y} : D \mapsto \mathbb{R}^n$) which has the value zero at all points of the boundary ∂D but is otherwise arbitrary, we form $y(\cdot, \alpha) = y_0 + \alpha Y$ (or $\underline{y}(\cdot, \alpha) = \underline{y}_0 + \alpha \underline{Y}$). Then for sufficiently small $\alpha, y(\cdot, \alpha)$ (or $\underline{y}(\cdot, \alpha)$) is an admissible function. We set

$$J(y(\cdot, \alpha)) = \Phi(\alpha) \quad (\text{or } J(\underline{y}(\cdot, \alpha)) = \Phi(\alpha)).$$

At the point $\alpha = 0$, Φ has a relative minimum since y_0 (or $\underline{y}_0$) is a solution of the variational problem. It follows that $\Phi'(0) = 0$. We will pursue this problem with one unknown function further and, after that, provide the formulae for the problem with several unknown functions

$$\Phi'(0) = \int_D \left[f_y^0(x_\beta) Y(x_\beta) + \sum_{\alpha=1}^{m} f_{y_\alpha}^0(x_\beta) Y_{x_\alpha}(x_\beta) \right] dx_1 \cdots dx_m, \tag{10.6}$$

where

$$f_y^0(x_\beta) = f_y(x_\beta, y_0(x_\beta), y_{0x_\alpha}(x_\beta)),$$
$$f_{y_\alpha}^0(x_\beta) = f_{y_\alpha}(x_\beta, y_0(x_\beta), y_{0x_\alpha}(x_\beta)).$$

If we now apply the Theorem of Gauss to the vector function $\underline{u}$ with components u_α,

$$u_\alpha(x_\beta) = f_{y_\alpha}^0(x_\beta) Y(x_\beta), \quad \alpha = 1, \ldots, m,$$

we obtain the formula

$$\begin{aligned}
&\int_D \sum_{\alpha=1}^{m} \frac{\partial}{\partial x_\alpha}[f^0_{y_\alpha}(x_\beta)Y(x_\beta)]\, dx_1 \cdots dx_m \\
&\quad = \int_D \sum_{\alpha=1}^{m} \left[\frac{\partial}{\partial x_\alpha} f^0_{y_\alpha}(x_\beta)Y(x_\beta) + f^0_{y_\alpha}(x_\beta)Y_{x_\alpha}(x_\beta)\right] dx_1 \cdots dx_m \\
&\quad = \int_D \sum_{\alpha=1}^{m} f^0_{y_\alpha}(p)Y(p)n_\alpha(p)\, d\sigma(p) = 0,
\end{aligned} \tag{10.7}$$

where the last equality follows from the fact that for all p on ∂D, $Y(p) = 0$. Thus we have

$$\Phi'(0) = \int_D \left[f^0_y(x_\beta) - \sum_{\alpha=1}^{m} \frac{\partial}{\partial x_\alpha} f^0_{y_\alpha}(x_\beta)\right] Y(x_\beta)\, dx_1 \cdots dx_m. \tag{10.8}$$

To enable us to obtain the *Euler differential equations*

$$f^0_y(x_\beta) - \sum_{\alpha=1}^{m} \frac{\partial}{\partial x_\alpha} f^0_{y_\alpha}(x_\beta) = 0 \quad \text{for all} \quad x_\beta \in D, \tag{10.9}$$

we will need the following important theorem.

(10.10) Fundamental Theorem of the Calculus of Variations (for the m-fold variational integral)
Let $D \subset \mathbb{R}^m$ be a region and let ψ be a continuous function defined on the closure $\overline{D} = D \bigcup \partial D$ of D. Suppose also that for all smooth functions Y on $\overline{D}$ which have the value zero on the boundary ∂D of D,

$$\int_D \psi(x_\beta)Y(x_\beta)\, dx_1 \cdots dx_m = 0. \tag{10.11}$$

Then ψ is identically zero on D.

Proof

Suppose there is a point $(x_{0\beta})$ in D for which $\psi(x_{0\beta}) \neq 0$. Then there is an m-dimensional sphere $K(x_0, r_1) = \{(x_\beta) \in \mathbb{R}^m : |(x_\beta) - x_{0\beta}| < r_1\}$ which lies entirely within D. Because of the continuity of ψ there is contained within $K(x_0, r_1)$ a sphere $K(x_0, r_2)$ with $r_2 \leq r_1$ so that for all points (x_β) in the sphere $K(x_0, r_2)$, the number $\psi(x_\beta)$ has the same sign as $\psi(x_{0\beta})$. The function $Y : D \mapsto \mathbb{R}$ defined by

$$Y(x_\beta) = \begin{cases} [r_2^2 - |(x_\beta) - (x_{0\beta})|^2]^4 & \text{if} \quad (x_\beta) \in K(x_0, r_2), \\ 0 \quad \text{if } (x_\beta) \in D, \text{ and } (x_\beta) \notin K(x_0, r_2), \end{cases}$$

is smooth on $\overline{D}$ and vanishes on ∂D. According to assumption, the integral (10.11) equals zero for this function Y. On the other hand, we have

$$\int_D \psi(x_\beta) Y(x_\beta)\, dx_1 \cdots dx_m = \int_{K(x_0, r_2)} \psi(x_\beta) Y(x_\beta)\, dx_1 \cdots dx_m$$

$$\begin{aligned} \text{which is} < 0 \quad &\text{if} \quad \psi(x_{0\beta}) < 0 \\ \text{or} > 0 \quad &\text{if} \quad \psi(x_{0\beta}) > 0. \end{aligned}$$

This is a contradiction; thus the assumption that there is a point $(x_{0\beta})$ in D for which $\psi(x_{0\beta}) \neq 0$ is false.

We provide here without comment the formulae analogous to (10.6), (10.8) and (10.9), but for the variational problem with multiple integrals and several unknown functions $y_1, y_2, \ldots, y_n$ or, equivalently, one vector function $\underline{y}$

$$\begin{aligned} \Phi'(0) = \int_D \sum_{i=1}^{n} \Big[& f^0_{y_i}(x_\beta) Y_i(x_\beta) \\ & + \sum_{\alpha=1}^{m} f^0_{y_{i\alpha}}(x_\beta) \frac{\partial Y_i}{\partial x_\alpha}(x_\beta) \Big]\, dx_1 \cdots dx_m, \end{aligned} \tag{10.6'}$$

$$\begin{aligned} \Phi'(0) = \sum_{i=1}^{n} \int_D \Big[& f^0_{y_i}(x_\beta) \\ & - \sum_{\alpha=1}^{m} \frac{\partial}{\partial x_\alpha} f^0_{y_{i\alpha}}(x_\beta) \Big]\, Y_i(x_\beta)\, dx_1 \cdots dx_m. \end{aligned} \tag{10.8'}$$

The *Euler equations* are

$$f^0_{y_i}(x_\beta) - \sum_{\alpha=1}^{m} \frac{\partial}{\partial x_\alpha} f^0_{y_{i\alpha}}(x_\beta) = 0; \quad i = 1, 2, \ldots, n. \qquad (10.9')$$

The Euler differential equations (10.9) and (10.9′) are normally partial differential equations. Their solutions are called *extremals.*

In the two examples (minimal surface and oscillation of a string) of Section 10.1, only the derivatives of the unknown functions appear explicitly in the integrands. The integrands depend, therefore, only on the last group of variables, the y_α, but not on the variables x_α and y. For this class of problem, the Euler differential equations have the form

$$\sum_{\alpha,\beta=1}^{m} f_{y_\alpha y_\beta}(x_\rho, y(x_\rho), y_{x_\sigma}(x_\rho)) y_{x_\alpha x_\beta}(x_\rho) = 0; \quad \sigma, \rho = 1, \ldots, m.$$

It can be seen immediately that every linear function y is an extremal. However, for $m > 1$, the linear functions are not the only extremals.

If in the example of the oscillating string (cf. (10.5)¡) the radical in the integrand is expanded in powers of y_x^2 (binomial formula), then one obtains for the integrand

$$f(x, t, y, y_x, y_t) = \tfrac{1}{2}[\rho y_t^2 - \mu y_x^2] + \text{'higher terms'}.$$

Since in the most important cases y_x is a small value, we may ignore these 'higher terms'. The Euler differential equation is then

$$\rho y_{tt} - \mu y_{xx} = 0,$$

which is the vibrating string equation.

Example. Equilibrium state oscillations of a membrane

A *membrane* is a two-dimensional surface which in its rest position is a plane elastic body. We want to study small deformations of the membrane. Let $z(x, y)$ denote the displacement of the membrane perpendicular to the xy-plane (rest position) at the point $(x, y) \in D$. The quantity z will be assumed small so that second and higher powers of z, z_x, z_y can be ignored. The potential energy U of a deformed membrane is proportional to the difference of the areas of the deformed membrane and of the membrane at rest

$$U = U(z) = \mu \int_D \left(\sqrt{1 + z_x^2(x,y) + z_y^2(x,y)} - 1\right)\, dx\, dy,$$

$$\mu = \text{ tension}, \quad \mu > 0.$$

For small values of z_x and z_y, a good approximation of $U(z)$ is given by

$$J(z) = \frac{\mu}{2} \int_D [z_x^2(x,y) + z_y^2(x,y)]\, dx\, dy. \tag{10.12}$$

Assume now that the displacement z on the boundary ∂D of D takes on prescribed fixed values $z(p) = \varphi(p)$ for $p \in \partial D$. Then in the equilibrium state the displacement z_0 at a point interior to D is determined so that the potential energy $J(z_0)$ is minimal

$$J(z) \geq J(z_0).$$

This is the principle of minimum potential energy. The Euler differential equation of the variational problem $J(z) \to \min$ is given by

$$z_{xx}(x,y) + z_{yy}(x,y) = \Delta z = 0,$$

which is the *Laplace differential equation*. Here Δ is the Laplacian operator $\frac{\partial^2}{\partial x^2} + \frac{\partial^2}{\partial y^2}$. As is well known, the Laplace equation arises in many problems of physics. For the most part these problems are related to a variational principle for which the variational integral is the *Dirichlet integral* (10.12).

We consider next the general case in which an external force is applied to each point of the membrane; for example, the gravitational force which depends on the surface density $k(x, y)$ (= force per unit of area at (x, y)).

If the force vector at each point is perpendicular to the rest position of the membrane, then its direction is given by the sign of the function k. Thus, $k(x,y) > 0$ if the force has the direction of the positive z-axis and $k(x,y) < 0$ if the force has the direction of the negative z-axis. Let the boundary of the membrane be fixed

$$z(p) = \phi(p) \quad \text{for} \quad p \in \partial D. \tag{10.13}$$

Then the expression for the potential energy will have an additional term so that

$$U(z) = \int_D \left[\frac{\mu}{2}(z_x^2(x,y) + z_y^2(x,y)) + k(x,y)z(x,y)\right] dx\, dy. \tag{10.14}$$

The Euler differential equation of this variational integral (10.14) is the *Poisson equation*

$$\mu\Delta z(x,y) = \mu(z_{xx}(x,y) + z_{yy}(x,y)) = k(x,y). \tag{10.15}$$

The solution z of the boundary value problem (10.13) and (10.15) yields the deformation of the membrane in the equilibrium position.

If the external force which acts on the membrane is time dependent so that the surface density k is a function of x, y and t, then we obtain the law of motion of the membrane from *Hamilton's principle*: the deformation z (dependent on x, y and t) is such that the action integral

$$J(z) = \int_{t_0}^{t_1} (T(t) - U(t))\, dt$$

is stationary. Here

$$T(t) = \frac{\rho}{2}\int_d z_t^2(x,y,t)\, dx\, dy$$

is the kinetic energy of the membrane at the time t, ρ is the (constant) mass per unit area, and

$$U(t) = \int_D \left[\frac{\mu}{2}(z_x^2 + z_y^2) + kz\right] dx\, dy$$

(arguments of z_x, z_y, k and z are x, y and t)

is the potential energy of the membrane at the time t. The Euler differential equation of the variational integral $J(z)$ is

$$\mu \Delta z(x,y,t) - \rho z_{tt}(x,y,t) - k(x,y,t) = 0.$$

This is the oscillation equation of the membrane; Δ is the Laplace operator.

Note

To obtain solutions of the different boundary value problems consisting of the above differential equations and their boundary conditions, one could also try to find a solution for the associated variational problem. Naturally, methods different from those we have introduced thus far must be used. These are the so-called direct methods which we will consider briefly in Chapter 12.

Exercise

For the variational problem with free boundary conditions (i.e. $y|\partial D$ may be an arbitrary differentiable function) and with a multiple variational integral

$$J(y) = \int_D f(x_\beta, y(x_\beta), y_{x_\alpha}(x_\beta))\, dx_1 \cdots dx_m,$$

derive the natural boundary condition

$$\sum_{\alpha=1}^{m} f_{y_\alpha}(p, y(p), y_{x_\beta}(p)) n_\alpha(p) = 0 \quad \text{for} \quad p \in \partial D, \tag{10.16}$$

where $n_\alpha(p)$ are the components of the external (outward) normal n of ∂D at p.

10.3 SUFFICIENT CONDITIONS

The necessary conditions for the variational problem with multiple integrals which correspond to the Legendre, Weierstrass and

Jacobi conditions are also known. We will not discuss them here. Rather, in this section we will derive the usual sufficient conditions for the fixed boundary problem with convex integrands and also show how, with the help of an invariant integral, a sufficient condition can be obtained. An example will show how the sufficient condition functions with an invariant integral.

We will set down the sufficiency condition for convex integrands for the case of several unknown functions $y_1, y_2, \ldots, y_n$; the case involving only one unknown function y_1 is included here (set $n = 1$ and drop the Latin indices $(i, j, \ldots)$).

(10.17) Theorem

Assume that the integrand f is defined for each fixed m-tuple $(x_\alpha) \in D$ on a convex set $\{(x_\alpha, y_i, y_{i\alpha}) \in$ domain of definition of $f\}$ and that, as a function of the $n + nm$ variables y_i, $y_{\alpha i}$ is convex. Suppose also that $\underline{y}^0 = (y_1^0, y_2^0, \ldots y_n^0) : D \mapsto \mathbb{R}^n$ is a two-fold smooth vector function which satisfies the boundary conditions of the fixed boundary problem as well as the Euler differential equations (10.9′). Then $\underline{y}^0$ is a solution of the fixed boundary problem.

Proof

Suppose $\underline{y} : D \mapsto \mathbb{R}^n$ is an admissible function. We need to consider the difference $J(\underline{y}) - J(\underline{y}^0)$. As in all theorems of this kind, we use the following inequality valid for convex functions:

$$f(x_\alpha, y_i, y_{i\alpha}) - f(x_\alpha, y_i^0, y_{i\alpha}^0) \geq \sum_{i=1}^{n} f_{y_i}(x_\alpha, y_j^0, y_{j\alpha}^0)(y_i - y_i^0)$$
$$+ \sum_{i=1}^{n} \sum_{\beta=1}^{m} f_{y_{i\beta}}(x_\alpha, y_j^0, y_{j\alpha}^0)(y_{i\beta} - y_{i\beta}^0). \text{ (cf. (3.9))}. \tag{10.18}$$

Using this inequality, we can now write

$$J(\underline{y}) - J(\underline{y}^0) = \int_D [f(x_\alpha, y_i(x_\alpha), y_{ix_\beta}(x_\alpha))$$
$$-f(x_\alpha, y_i^0(x_\alpha), y_{ix_\beta}^0(x_\alpha))]\, dx_1 \cdots dx_m$$

$$\geq \int_D \left[\sum_{i=1}^{n} f^0_{y_i}(x_\alpha)(y_i(x_\alpha) - y^0_i(x_\alpha)) \right.$$
$$\left. + \sum_{i=1}^{n} \sum_{\beta=1}^{m} f^0_{y_{i\beta}}(x_\alpha) \frac{\partial}{\partial x_\beta}(y_i - y^0_i)(x_\alpha) \right] dx_1 \cdots dx_m.$$

We have used here the abbreviated notation

$$f^0_{y_i}(x_\alpha) = f_{y_i}(x_\alpha, y^0_j(x_\alpha), y^0_{jx_\beta}(x_\alpha)),$$
$$f^0_{y_{i\beta}}(x_\alpha) = f_{y_{i\beta}}(x_\alpha, y^0_j(x_\alpha), y^0_{jx_\beta}(x_\alpha)).$$

With the help of the integral theorem of Gauss (10.3) the double sum can be integrated by parts. The boundary integral which arises in this process vanishes since for the fixed boundary problem $y_i(p) = y^0_i(p)$ for $i = 1, 2, \ldots, n$ and for all $p \in \partial D$. The remaining part, therefore, is

$$J(\underline{y}) - J(\underline{y}^0) \geq \int_D \left[\sum_{i=1}^{n} f^0_{y_i}(x_\alpha) - \sum_{\beta=1}^{m} \frac{\partial}{\partial x_\beta} f^0_{y_{i\beta}}(x_\alpha) \right]$$
$$\cdot (y_i(x_\alpha) - y^0_i(x_\alpha))\, dx_1 \cdots dx_m = 0,$$

where y^0 satisfies the Euler differential equation.

Since the integrand of the Dirichlet integral (10.12) is convex, we can apply Theorem (10.17) to it. Furthermore, the integrand of the variational integral (10.14) (equilibrium state of a membrane) is also convex.

We now come to the determination of sufficient conditions for the fixed boundary problem achieved with the help of a path-independent integral. In the discussion of the Hilbert integral, whose path-independence was shown in Section 6.4, we spoke of a 'potential' P whose differential dP was the integrand of the invariant integral. In the case of the multiple integrals we have in place of P an $(m-1)$-differential form ψ on the $(x_1, x_2, \cdots, x_m,\ y_1, y_2, \cdots, y_n)$-space, and in place of dP the differential $d\psi$ of ψ. We will need the generalized Stoke's theorem

$$\int_{\mathbb{F}} d\psi = \int_{\partial \mathbb{F}} d\psi;$$

here integration is over an m-dimensional differentiable surface $\mathbb{F}$ in $x_\alpha y_i$-space or, respectively, over the smooth boundary $\partial\mathbb{F}$ of $\mathbb{F}$. If $\mathbb{F}_1$ and $\mathbb{F}_2$ are m-dimensional surfaces with the same boundary $\partial\mathbb{F}_1 = \partial\mathbb{F}_2$, then

$$\int_{\mathbb{F}_1} d\psi = \int_{\mathbb{F}_2} d\psi.$$

That is, $\int_{\mathbb{F}} d\psi$ is a 'path-independent' integral. Each vector function $\underline{y} : D \subset \mathbb{R}^m \mapsto \mathbb{R}^n$ defines an m-dimensional surface $\mathbb{F}_{\underline{y}}$ in (x_α, y_i)-space for which

$$(x_\alpha) \in D \mapsto (x_\alpha, y_i(x_\alpha)), \quad i = 1, 2, \ldots, n, \quad \alpha = 1, \ldots, m,$$

is a parametric representation. For every m-differential form ω defined on a region of the $x_\alpha y_i$-space there is a corresponding function Ω (dependent on the variables $x_\alpha, y_i, y_{i\alpha}$) with whose help the integration of ω over a surface of type $\mathbb{F}_{\underline{y}}$ can be achieved

$$\int_{\mathbb{F}_{\underline{y}}} \omega = \int_D \Omega(x_\alpha, y_i(x_\alpha), y_{ix_\beta}(x_\alpha))\, dx_1 \cdots dx_m. \tag{10.19}$$

With this preparation we can now formulate a sufficient condition

(10.20) Theorem

Let $\underline{y}_0 = (y_i^0) = (y_1^0, y_2^0, \cdots, y_n^0) : D \mapsto \mathbb{R}^n$ *be a two-fold smooth vector function which satisfies the boundary conditions of the previously stated fixed boundary problem. If for* $\underline{y}^0$ *there is an* m*-differential form* ω *which is the differential* $d\psi$ *of an* $(m-1)$*-differential form* ψ *and for which the corresponding function* ω *has the properties*

$$f(x_\alpha, y_i^0(x_\alpha), y_{ix_\beta}^0(x_\alpha)) - \Omega(x_\alpha, y_i^0(x_\alpha), y_{ix_\beta}^0(x_\alpha)) = 0 \quad \text{for all } (x_\alpha) \in D \tag{10.21}$$

and

$$f(x_\alpha, y_i, y_{i\alpha}) - \Omega(x_\alpha, y_i, y_{i\alpha}) \geq 0 \tag{10.22}$$

for all $(x_\alpha, y_i, y_{i\alpha})$ *of the domains of definition of* f *and* Ω, *then* $\underline{y}^0$ *is a solution of the variational problem.*

Proof

The integral over Ω depends only on the values of the function on the boundary ∂D of D and so is a path-independent integral. For all admissible vector functions $y : D \mapsto \mathbb{R}^n$ the integral over Ω must therefore provide the same value. This value, however, agrees with the value of the integral $J(\underline{y}^0)$ everywhere because of (10.21).

We want to determine next the difference $J(\underline{y}) - J(\underline{y}^0)$ for an arbitrary admissible vector function $\underline{y}$

$$\begin{aligned} J(\underline{y}) - J(\underline{y}^0) &= J(\underline{y}) - \int_D \Omega(x_\alpha, \overset{0}{y_i}(x_\alpha), \overset{0}{y}_{ix_\beta}(x_\alpha))\, dx_1 \cdots dx_m \\ &= J(y) - \int_D \Omega(x_\alpha, y_i(x_\alpha), y_{ix_\beta}(x_\alpha))\, dx_1 \cdots dx_m \\ &= \int_D [f(x_\alpha, y_i(x_\alpha), y_{ix_\beta}(x_a)) \\ &\quad -\Omega(x_\alpha, y_i(x_\alpha), y_{ix_\beta}(x_\alpha))]\, dx_1 \cdots dx_m \geq 0. \end{aligned}$$

(Inequality (10.22) has been used here.) As a consequence, $J(\underline{y}) \geq J(\underline{y}_0)$ for all admissible functions $\underline{y}$. We conclude that $\underline{y}^0$ is therefore a solution of the variational problem with multiple integrals.

Example

Consider the following variational problem. Among all smooth functions z which are defined on the closure of the region D in the xy-plane, where D has a smooth boundary ∂D and where

$$z(p) = 0 \quad \text{for all} \quad p \in \partial D, \tag{10.23}$$

find those functions for which the integral

$$\begin{aligned} J(z) &= \int_D [A(x,y) z^2(x,y) + z_x^2(x,y) + z_y^2(x,y)]\, dx\, dy \\ &= \int_D f(x, y, z(x,y) z_x(x,y) z_y(x,y))\, dx\, dy \end{aligned}$$

has a minimal value.

The Euler differential equation of the problem is

$$A(x,y)z(x,y) - z_{xx}(x,y) = z_{yy}(x,y) = 0 \qquad (10.24)$$
$$\text{(more concisely: } \Delta z - Az = 0).$$

The function $z_0 = 0$ is a solution of the boundary value problem (10.23) and (10.24). If the function A takes on only non-negative values, then the integrand is convex so that Theorem (10.17) is applicable. If A fails to be non-negative, then the method of invariant integrals – with the extremal $z_0 = 0$ – may be tested on this simple example. We will choose here a substitution for the invariant integral which is due to Th. de Donder and H. Weyl (cf. [10]). Accordingly, set

$$\psi(x,y,z) = S_1(x,y,z)\,dy - S_2(x,y,z)\,dx.$$

Here S_1 and S_2 are yet to be determined (differentiable) functions on a subset of the xyz-space

$$\omega = d\psi = dS_1 \wedge dy - dS_2 \wedge dx.$$

It follows that the integrand Ω which belongs to the differential form ω has the form

$$\begin{aligned}\Omega(x,y,z,z_x,z_y) = {} & S_{1x}(x,y,z) + S_{2y}(x,y,z)\\ & + S_{1z}(x,y,z)z_x + S_{2z}(x,y,z)z_y.\end{aligned}$$

The conditions of Theorem (10.20) are

$$\begin{aligned} f(x,y,0,0,0) = 0 = {} & \Omega(x,y,0,0,0)\\ = {} & S_{1x}(x,y,0) + S_{2y}(x,y,0) \quad \text{for all } (x,y) \in D, \end{aligned} \qquad (10.21')$$

and

$$\begin{aligned} f(x,y,z,z_x,z_y) - \Omega(x,y,z,z_x,z_y) = {} & Az^2 + z_x^2 + z_y^2\\ & - S_{1x} - S_{2y} - S_{1z}z_x - S_{2z}z_y \geq 0 \end{aligned} \qquad (10.22')$$

for all x, y, z, z_x, z_y in the domains of definition of f and Ω. These are relatively weak conditions which allow many possibilities for the construction of the functions S_1 and S_2 but include few instructions. So let us sharpen condition (10.21′) somewhat by requiring that there be defined on the xyz-space a slope function (p_1, p_2) so that for all (x, y, z)

$$\begin{aligned} f(x, y, z, p_1(x, y, z), &p_2(x, y, z)) \\ &- \Omega(x, y, z, p_1(x, y, z), p_2(x, y, z)) = 0. \end{aligned} \tag{10.21''}$$

If (10.22′) is compared with (10.21″), it can be seen that for each (x, y, z) the function $f - \Omega$ (dependent on z_x and z_y) has a minimum at the point $z_x = p_1(x, y, z), z_y = p_2(x, y, z)$. As a consequence, the partial derivatives of $f - \Omega$ with respect to z_x and z_y must vanish at this point:

$$\begin{aligned} f_{z_x}(\cdots) = \Omega_{z_x}(\cdots), \quad 2p_1(x, y, z) = S_{1z}(x, y, z), \\ f_{z_y}(\cdots) = \Omega_{z_y}(\cdots), \quad 2p_2(x, y, z) = S_{2z}(x, y, z). \end{aligned} \tag{10.25}$$

Along the extremals z_0, p_x and p_y must agree with z_{0x} and z_{0y}, respectively. Consequently, the following conditions must hold for the functions S_1 and S_2:

$$\begin{aligned} p_1(x, y, 0) = z_{0x}(x, y) = 0 = S_{1z}(x, y, 0), \\ p_2(x, y, 0) = z_{0y}(x, y) = 0 = S_{2z}(x, y, 0). \end{aligned} \tag{10.26}$$

If we now substitute (10.25) into (10.21″), we obtain a partial differential equation for the two functions S_1 and S_2, the so-called *Hamilton–Jacobi differential equations*, which, for our example, are

$$\begin{aligned} A(x, y)z^2 &+ \tfrac{1}{4}(S_{1z})^2 + \tfrac{1}{4}(S_{2z})^2 \\ &- [S_{1x} + S_{2y} + \tfrac{1}{2}(S_{1z})^2 + \tfrac{1}{2}(S_{2z})^2] \\ = A(x, y)z^2 &- S_{1x} - S_{2y} - \tfrac{1}{4}[(S_{1z})^2 + (S_{2z})^2] = 0. \end{aligned} \tag{10.27}$$

(Arguments for the partial derivatives of S are x, y, z.)

We have now *one* partial differential equation for *two* unknown functions so that we can prescribe one function and thereby fulfil (10.26) and (10.22′).

Suppose we now set $S_2 = 0$ (cf. (10.26)). In order to be able to proceed further, we must know the function A. We examine the case where $A(x, y) = -A_0 =$ constant, $A_0 > 0$. In accord with (10.26) we choose

$$S_1(x, y, z) = s(x)z^2,$$

where s is a yet to be determined function. Equation (10.27) now reduces to

$$Az^2 - s'(x)z^2 - s^2(x)z^2 = -z^2(A_0 + s'(x) + s^2(x)) = 0, \quad (10.28)$$

which is satisfied only if

$$A_0 + s'(x) + s^2(x) = 0.$$

This (ordinary) differential equation can be solved by separation of variables. Its solution is

$$s(x) = -\sqrt{A_0}\tan\left[\sqrt{A_0}(x + k)\right]$$

$$(k = \text{ is the constant of integration}),$$

which is defined for $x \in (-\frac{\pi}{2c} - k, \frac{\pi}{2c} - k)$, $c = \sqrt{A_0}$.

The functions S_1 and S_2 are now determined. It remains to check the condition (10.22′)

$$\begin{aligned} &Az^2 + z_x^2 + z_y^2 - S_{1x} - S_{2y} - S_{1z}z_x - S_{2z}z_y \\ = \quad & -A_0z^2 + z_x^2 + z_y^2 - s'(x)z^2 - 2s(x)zz_x \\ = \quad & (z_x - s(x)z)^2 + z_y^2 - z^2(A_0 + s'(x) + s^2(x)) \\ = \quad & (z_x - s(x)z)^2 + z_y^2 \geq 0. \end{aligned} \quad (10.28)$$

We summarize the result as follows: The variation problem

$$\int_D (-A_0z^2(x, y) + z_x^2(x, y) + z_y^2(x, y))\, dx\, dy \to \min,$$

boundary conditions $z(p) = 0$ for all $p \in \partial D$,

has the solution $z_0 = 0$ provided the domain of definition $\overline{D} = D \bigcup \partial D$ of the admissible functions lies in a strip of the xy-plane

$$\overline{D} \subset \{(x,y) \in \mathbb{R}^2 : -\frac{\pi}{2\sqrt{A_0}} - k < x < \frac{\pi}{2\sqrt{A_0}} - k, y \in \mathbb{R}\},$$

where k is an arbitrary real number.

The last-named requirement on the domain of definition $\overline{D}$ is necessary since without it the function S_1 would not be defined. The question naturally arises whether $z_0 = 0$ remains a solution of the variational problem for a larger domain of definition $\overline{D}$. To investigate this problem would require the use of Jacobi's necessary condition which we have not discussed for this problem.

Exercises

(10.29) Find the Euler differential equation for the variational problem for the minimal surface

$$J(z) = \int_D \sqrt{1 + z_x^2(x,y) + z_y^2(x,y)}\, dx\, dy \to \min$$

(plus boundary conditions)

and show that for the fixed boundary problem every extremal which satisfies the boundary conditions is a solution of the variational problem.

(10.30) Let D be the rectangle $D = \{(x,y) \in \mathbb{R}^2 : 0 \leq x, y \leq \pi\}$. Among all smooth functions $z : D \mapsto \mathbb{R}$ which vanish on the boundary ∂D of the rectangle, find a function z_0 for which the integral

$$\int_D [z_x^2(x,y) + z_y^2(x,y) - z(x,y) \sin x \sin y]\, dx\, dy$$

has its smallest value. It may be assumed here that all the theorems which we have proved are valid when the domain of definition is the rectangle D in the same way they are valid for the domain D of the theorems, since the Gauss integral of Theorem (10.3) is also valid for rectangles. Show also that z_0, defined by

$$z_0(x, y) = \tfrac{1}{2} \sin x \sin y,$$

is an extremal of the variational problem and that it satisfies the boundary conditions. Why is z_0 a solution of the problem? Use an invariant integral to test the sufficiency condition for this problem as was done in the example given above. In so doing, functions S_1 and S_2 of the type

$$S_\alpha(x, y, z) = R_\alpha(x, y)z + T_\alpha(x, y), \quad \alpha = 1, 2,$$

can be employed.

11

Variational problems with side conditions

There are problems for which conditions other than boundary conditions are imposed on admissible functions or curves. One example is variational problem (1.4) for the closed curve of fixed length, which encloses the largest plane area. Further examples deal with the motion of a pendulum and the problem of the equilibrium state of a heavy chain.

(11.1) Motion of a mathematical pendulum

Let a point-mass m be attached to a string of length L whose mass may be ignored (see Fig. 11.1). The other end of the string is fastened securely at the point $A = (0, L)$.

We want to determine the path $(x(t), y(t))$ of the point-mass under the assumption that during the motion the string remains taut. We assume also that at time $t = 0$ the point-mass is displaced to a certain initial position (x_0, y_0) and that it has an initial velocity (v_1, v_2). According to Hamilton's principle, the mass moves so that the integral

$$J(x, y) = \int_0^{t_1} [T(t) - U(t)]\, dt \qquad (11.2)$$

is stationary. Here $T(t)$ denotes the kinetic energy and $U(t)$ the potential energy of the pendulum at time t. Admissible func-

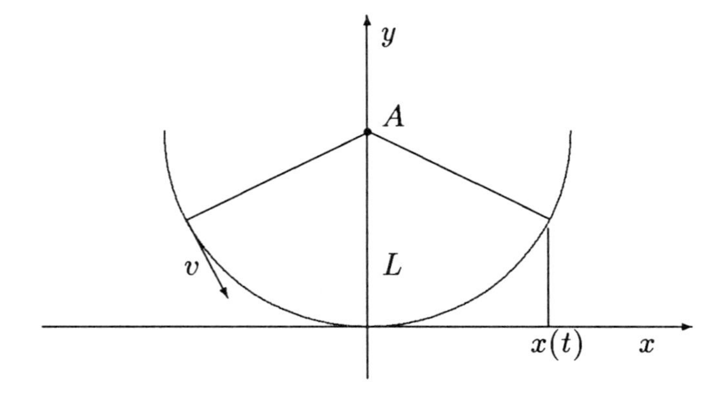

Fig. 11.1. The mathematical pendulum.

tions are the piecewise smooth vector functions (x, y) with the time t as the independent variable and which satisfy the initial conditions

$$x(0) = x_0, \quad y(0) = y_0,$$
$$x'(0) = v_1, \quad y'(0) = v_2.$$

A further condition requires that for all t the points $(x(t), y(t))$ lie on a circle with centre at A and of radius L

$$x^2(t) + (y(t) - L)^2 = L^2. \tag{11.3}$$

(11.4) Equilibrium position of a chain

A heavy homogeneous chain of length L is supported at its ends. In its equilibrium position, therefore, it hangs under the influence of gravity in a vertical plane. We want to find the curve k in this plane which describes the position of the chain. This problem can be formulated in various ways as a variational problem with side conditions.

1. Let $A = (a, y_a)$ and $B = (b, y_b)$, $a < b$, be the coordinates of the endpoints of the chain in a Cartesian coordinate system whose negative axis has the direction of the force of gravity.

The equilibrium state of the chain is achieved when its centre of gravity is at its lowest point, i.e. when the potential energy of the chain is smallest. It is evident that the solution may be found among those curves which can be expressed in terms of a parameter x; i.e. among curves having a parametric representation

$$x \mapsto (x, y(x)), \quad a \leq x \leq b.$$

The variational integral is now given by

$$J(y) = g \int_0^L y(x)\rho \, ds = g\rho \int_a^b y(x)\sqrt{1 + y'^2(x)} \, dx, \tag{11.5}$$

where ρ denotes the mass of the chain per unit length, g the gravitational constant, and s the length of the curve. The side condition on the admissible functions has the form

$$\begin{aligned} \text{length of the chain} = L &= \text{constant} \\ &= \int_a^b \sqrt{1 + y'^2(x)} \, dx. \end{aligned} \tag{11.6}$$

The boundary conditions are $y(a) = y_a$ and $y(b) = y_b$. In order that there be admissible curves, of course, it is necessary that the chain be long enough: $L^2 \geq (y_b - y_a)^2 + (b - a)^2$.

2. If we use s, the arc length, as a parameter instead of x, then the parametric representation becomes

$$s \mapsto (x(s), y(s)), \quad 0 \leq s \leq L,$$

and the variational integral has the form

$$J(x, y) = g\rho \int_0^L y(s) \, ds \to \min. \tag{11.7}$$

The boundary conditions for this problem are: $(x(0), y(0)) = (a, y_a)$, $(x(L), y(L)) = (b, y_b)$, and the side condition is

$$x'^2(s) + y'^2(s) = 1 \quad \text{for all } s \in (0, L). \tag{11.8}$$

Clearly, s, the arc length parameter, is an excellent one to use here.

In these examples we have presented the most important types of side conditions: (11.3) is an algebraic equation which must be satisfied by the desired function while (11.8) is a differential equation to be satisfied. The side condition (11.6) has the form

$$J^*(y) = \text{constant} = \int_a^b f^*(x, y(x), y'(x))\, dx. \tag{11.9}$$

This kind of side condition is called (after Example (1.4)) the isoperimetric side condition. In this chapter we will derive necessary conditions for solutions of variational problems with the three types of side conditions mentioned here. These necessary conditions correspond to the Euler differential equations.

11.1 THE LAGRANGE MULTIPLIER RULE

Before we began to derive the Euler differential equation in Chapter 2, we looked at the simplest extreme value problem in which the relative minimum of a differentiable function Φ of one variable x was sought. In a similar way, we will proceed here to show how to determine a relative minimum of a differentiable function Φ of two variables x and y, which are not free variables but rather coupled to each other by a side condition. Here is an example: Find the relative minimum of the function $\Phi(x, y) = x^2 - y^2$ under the side condition $g(x, y) = x^2 + y^2 - 1 = 0$. The side condition is the equation of the unit circle. A minimum of Φ under the side condition $g(x, y) = 0$ occurs, therefore, at a point (x_0, y_0) on the unit circle where the value of Φ is not surpassed by any other Φ-values on the circle

$$\Phi(x_0, y_0) \leq \Phi(x, y) \quad \text{for all points } (x, y) \text{ of the unit circle.}$$

Whether or where Φ as a function on the xy-plane has a minimum is of no interest here.

Suppose, therefore, that Φ and g are smooth functions defined on an open set G in the xy-plane. For the following, we will

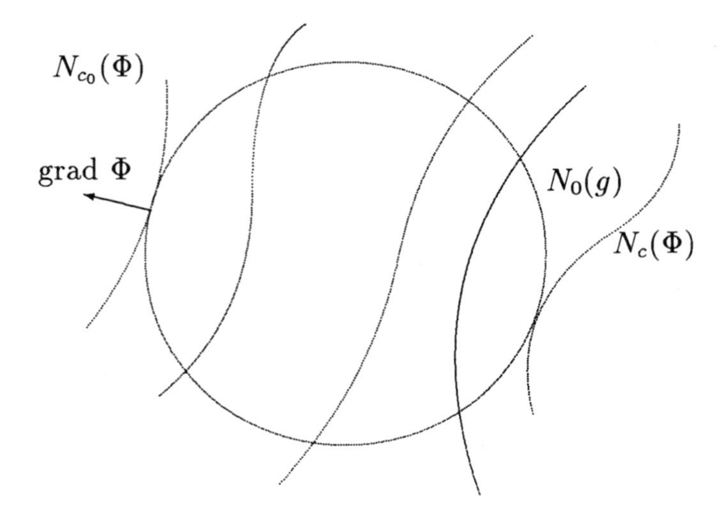

Fig. 11.2. Level curves.

assume that the equation $g(x, y) = 0$ can always be satisfied and that it is always locally solvable for x or for y. Such is the case if, for all $(x, y) \in G$,

$$\text{grad } g(x, y) = (g_x(x, y), g_y(x, y)) \neq (0, 0). \qquad (11.10)$$

Then the set $N_0(g) = \{(x, y) \in G : g(x, y) = 0\}$ of points which satisfy the side conditions forms the path of a curve. The Lagrange multiplier rule provides a necessary condition for a relative minimum (x_0, y_0) for Φ under the condition $g(x, y) = 0$. It depends on the following idea: Consider the various level curves (contour lines) of Φ (cf. Fig. 11.2)

$$N_c(\Phi) = \{(x, y) \in G : \Phi(x, y) = c = \text{constant}\},$$

and the contour line $N_0(g) = \{(x, y) \in G : g(x, y) = 0\}$. Let $c_0 = \Phi(x_0, y_0)$.

If the contour line $N_{c_0}(\Phi)$ has tangent at the point (x_0, y_0), then this tangent must agree with the tangent of $N_0(g)$ at (x_0, y_0); the contour lines touch each other. It follows that the normal vectors of these two tangents, i.e. grad $g(x_0, y_0)$ and grad $\Phi(x_0, y_0)$, must be multiples of each other

$$\text{grad } \Phi(x_0, y_0) = \lambda \,\text{grad } g(x_0, y_0). \qquad (11.11)$$

The factor λ is called the *Lagrange multiplier.*

(11.12) Theorem (Lagrange multiplier rule)
Let Φ and g be smooth functions on an open set G in the xy-plane. If (x_0, y_0) provides a relative minimum for Φ under the side condition $g(x, y) = 0$ and if the gradient $\operatorname{grad} g(x_0, y_0)$ *of g at the point (x_0, y_0) is not the zero vector, then there is a real number λ such that*

$$\operatorname{grad}(\Phi - \lambda g)(x_0, y_0) = 0.$$

If, therefore, we seek a relative minimum for a function Φ under the side condition $g = 0$, then the necessary condition takes the form

$$\Phi_x - \lambda g_x = 0, \quad \Phi_y - \lambda g_y = 0.$$

A relative minimum must fulfil the side condition $g = 0$ also. All told, therefore, we have three (in general, non-linear) equations for the three unknowns x_0, y_0 and λ. For the example $\Phi(x, y) = x^2 - y^2, g(x, y) = x^2 + y^2 - 1$, all of the assumptions of the theorem are satisfied, for $\operatorname{grad} g(x, y) = (2x, 2y)$ is the null vector only if $x = y = 0$. But the point $(0, 0)$ does not satisfy the condition $g(x, y) = 0$. The three equations for the determination of x_0, y_0 and λ are, therefore,

$$\begin{aligned}
2x - 2\lambda x &= 2x(1 - \lambda) = 0, \\
-2y - 2\lambda y &= 2y(-1 - \lambda) = 0, \\
x^2 + y^2 - 1 &= 0.
\end{aligned}$$

From the first equation it follows that either $x = 0$ or $\lambda = 1$. For $x = 0$ the third equation yields $y = \pm 1$. The function's value at the points $(0, 1)$ and $(0, -1)$ is $\Phi(0, 1) = \Phi(0, -1) = -1$. For $\lambda = 1$, it follows from the second equation that $y = 0$ so that, with the help of the side condition, $x = \pm 1$. Then the function's values are $\Phi(1, 0) = 1 = \Phi(-1, 0)$. On the closed and bounded set of points of the unit circle, the continuous function Φ must have a minimum. Thus the points $(0, 1)$ and $(0, -1)$ provide the desired minima. The condition (11.11) of Theorem (11.12) is also a necessary condition for maximum and stationary points. The points $(1, 0)$ and $(-1, 0)$ yield maximum values for $\Phi(x, y) = x^2 - y^2$ under the side condition $g(x, y) = x^2 + y^2 - 1 = 0$.

11.2 THE ISOPERIMETRIC PROBLEM

An *isoperimetric problem* is understood to be an extremal problem

$$J(y) = \int_a^b f(x, y(x), y'(x))\, dx \to \min,$$

for which the admissible functions are smooth functions on $[a, b]$ which, beside the boundary conditions $y(a) = y_a$ and $y(b) = y_b$, also satisfy a side condition of the form

$$J^*(y) = \int_a^b f^*(x, y(x), y'(x))\, dx = \text{constant} = c$$

(the isoperimetric side condition). Thus we impose on the integrand f^* of J^* the same assumptions as were imposed on the integrand f; i.e. F^* shall be three-fold smooth (partially) with respect to its variables x, y and y'. In this section we will apply the Lagrange multiplier rule to the isoperimetric problem and obtain a necessary condition. We assume that the conditions we have imposed on the admissible functions are satisfied and that the admissible function y_0, a two-fold smooth function, is a solution of the isoperimetric problem.

Let the function y_0 be changed now. Thus, choose two functions Y_1 and Y_2, defined and smooth on $[a, b]$. Also require that Y_1 and Y_2 satisfy the boundary conditions

$$Y_1(a) = Y_2(a) = Y_1(b) = Y_2(b) = 0; \qquad (11.13)$$

otherwise they shall be arbitrary. We can now form with these two functions a two-parameter family of functions $\hat{y}(\cdot, \alpha_1, \alpha_2)$,

$$\hat{y}(x, \alpha_1, \alpha_2) = y_0(x) + \alpha_1 Y_1(x) + \alpha_2 Y_2(x),$$

not all of which will be admissible functions. If we substitute the functions of the family $\hat{y}$ into the integrals J and J^*, we obtain the functions Φ and g

$$\begin{aligned}\Phi(\alpha_1,\alpha_2) &= J(\hat{y}(\cdot,\alpha_1,\alpha_2)) \\ &= \int_a^b f(x,\hat{y}(x,\alpha_1,\alpha_2),\ \hat{y}_x(x,\alpha_1,\alpha_2))\,dx,\end{aligned} \tag{11.14}$$

$$\begin{aligned}c+g(\alpha_1,\alpha_2) &= J^*(\hat{y}(\cdot,\alpha_1,\alpha_2)) \\ &= \int_a^b f^*(x,\hat{y}(x,\alpha_1,\alpha_2),\hat{y}_x(x,\alpha_1,\alpha_2))\,dx.\end{aligned} \tag{11.15}$$

These functions Φ and g are at least defined for all (α_1,α_2) which lie within a circle of sufficiently small radius about $(0,0)$. Within this circle both are partially differentiable with respect to α_1 and α_2. Here c is the constant which appears in the isoperimetric condition (11.9). Now $g(\alpha_1,\alpha_2)=0$ implies that $J^*(\hat{y}(\cdot,\alpha_1,\alpha_2))=c$. Furthermore, (11.13) means that $\hat{y}(\cdot,\alpha_1,\alpha_2)$ satisfies the boundary conditions. Thus $\hat{y}(\cdot,\alpha_1,\alpha_2)$ is an admissible function for the isoperimetric problem. By assumption, $y_0=\hat{y}(\cdot,0,0)$ is a solution of the isoperimetric problem. Consequently, $(\alpha_1,\alpha_2)=(0,0)$ provides a minimum for Φ under the side condition $g(\alpha_1,\alpha_2)=0$. In order to be able to apply Theorem (11.12), we must also check the gradient of g at the point $(0,0)$

$$\begin{aligned}g_{\alpha_1}(0,0) &= \int_a^b [f_y^*(x,y_0(x),y_0'(x))Y_1(x) \\ &\qquad + f_{y'}^*(x,y_0(x),y_0'(x))Y_1'(x)]\,dx \\ &= \int_a^b \left[f_y^{*0}(x)-\frac{d}{dx}f_{y'}^{*0}(x)\right]Y_1(x)\,dx.\end{aligned} \tag{11.16}$$

We have used integration by parts to integrate the second term here. Analogously,

$$g_{\alpha_2}(0,0)=\int_a^b [\cdots]Y_2(x)\,dx, \tag{11.17}$$

where the expression in the square brackets is the same as in the integrand for $g_{\alpha_1}(0,0)$.

From the explicit expression for $g_{\alpha_1}(0,0)$ and $g_{\alpha_2}(0,0)$ we can infer that if y_0 is an extremal of the integral J^*, then the expression contained in the square brackets in (11.16) vanishes for all $x \in [a,b]$ so that

$$g_{\alpha_1}(0,0) = g_{\alpha_2}(0,0) = 0; \qquad \text{grad}\, g(0,0) = (0,0).$$

Clearly, Theorem (11.12) is not applicable in this instance. Suppose, however, that y_0 is not an extremal for the integral J^*. Then the quantity in the square brackets will not be equal to zero for all $x \in [a,b]$, and there will be a suitable function Y_1 such that $g_{\alpha_1}(0,0) \neq 0$ independently of how Y_2 is chosen. Thus, suppose Y_1 is such a function, which we now hold fixed. The assumptions of Theorem (11.12) are satisfied and there is a real number λ, a Lagrange multiplier, for which

$$\text{grad}(\Phi - \lambda g)(0,0) = 0.$$

Thus,

$$\Phi_{\alpha_1}(0,0) - \lambda g_{\alpha_1}(0,0) = 0$$

and

$$\Phi_{\alpha_2}(0,0) - \lambda g_{\alpha_2}(0,0) = 0.$$

The multiplier may, by the way, be determined from the first of these equations, since $g_{\alpha_1}(0,0) \neq 0$. It remains to be determined whether the necessary condition

$$\Phi_{\alpha_2}(0,0) - \lambda g_{\alpha_2}(0,0) = \int_a^b \Big\{[f_y^0(x) - \lambda f_y^{*0}(x)] - \frac{d}{dx}[f_{y'}^0 - \lambda f_{y'}^{*0}](x)\Big\} Y_2(x)\, dx = 0$$

is satisfied for *all* smooth functions Y_2 with the boundary conditions $Y_2(a) = Y_2(b) = 0$. It is easily shown that the assumptions of the Fundamental Lemma of the Calculus of Variations (cf. (10.10) with $m = 1$) are satisfied. We have thus proved the following theorem.

(11.18) Theorem

Let the isoperimetric problem with the variational integral $J(y)$ and the isoperimetric side condition

$$J^*(y) = \int_a^b f^*(x, y(x), y'(x))\, dx = c$$

be given. Furthermore, let y_0, a two-fold smooth admissible function, be a solution of the isoperimetric problem but not an extremal of the variational problem with the variational integral J^*. Then there is a real number λ (Lagrange multiplier) such that y_0 is an extremal of a variational problem with the integrand $f - \lambda f^*$.

We can test this theorem by considering the example (11.4) of the equilibrium state of a homogeneous chain. The first of the two given mathematical models of the problem poses an isoperimetric problem

$$J(y) = g\rho \int_a^b y(x)\sqrt{1 + y'^2(x)}\, dx \to \min,$$

$$\text{boundary condition: } y(a) = y_a,\ y(b) = y_b.$$

In order that the conditions be fulfilled, the length of the chain must be greater than the distance between the chain's two support points; i.e.

$$L^2 \geq (y_b - y_a)^2 + (b - a)^2.$$

The extremals of the variational integral J^* are the functions $y(x) = k_1 x + k_2$, k_1 and k_2 constants. The extremal y^* of J^* which satisfies the boundary conditions of the isoperimetric problem represents the path joining the endpoints of the chain. For this function $J^*(y^*) = L^* = \sqrt{(y_b - y_a)^2 + (b - a)^2}$. In the event $L > L^*$, an admissible function of the isoperimetric problem cannot be an extremal of the variational integral J^*. A two-fold smooth differentiable function y_0 must then be an extremal of a variational integral with the integrand

$$f(x, y, y') - \lambda f^*(x, y, y') = g\rho y\sqrt{1 + y'^2} - \lambda\sqrt{1 + y'^2}.$$

Since this integrand is independent of x, we can solve the Euler differential equation with the procedure of Section 2.2

$$(g\rho y(x) - \lambda)\left[\frac{y'(x)}{\sqrt{1+y'^2(x)}} - \sqrt{1+y'^2(x)}\right] = c_1 = \text{constant}.$$

From this we obtain

$$-(g\rho y(x) - \lambda) = c_1\sqrt{1+y'^2(x)}.$$

The solutions of this differential equation are

$$y(x) = \frac{\lambda}{g\rho} = \text{constant if } c_1 = 0 \quad (\text{excluded if } L > L^*)$$

and

$$y(x) = c_0 + k\cosh\frac{1}{k}(x - x_0) \quad \text{if } c_1 \neq 0.$$

In the second solution, $c_0 = \frac{\lambda}{g\rho}$, $k = \frac{c_1}{g\rho}$, and x_0 is a constant of integration. (It is not surprising that curves represented by the function y are called 'catenaries' or 'chain lines'.)

Three yet to be determined constants, λ, c and x_0, appear in the expression for the extremals. These constants can be determined from the boundary conditions and the isoperimetric conditions

$$\begin{aligned} L &= J^*(y) \\ &= \int_a^b \sqrt{1+y'^2(x)}\,dx \\ &= \int_a^b \cosh\frac{1}{k}(x-x_0)\,dx \\ &= \left[k\sinh\frac{1}{k}(x-x_0)\right]_a^b. \end{aligned}$$

Exercise

Among all plane curves of length L which, with the exception of their endpoints, lie in the upper half-plane $y > 0$ of the xy-plane and have endpoints $A = (a, 0)$ and $B = (b, 0)$, find those

which together with the line segment joining A and B bound the largest area ($L > b - a > 0$). Solve the problem under the restriction that the admissible curve can be represented by a smooth function $y : [a, b] \mapsto \mathbb{R}$.

11.3 VARIATIONAL PROBLEMS WITH AN EQUATION AS A SIDE CONDITION

If the unknown functions of a variational problem are coupled to each other in the sense that each (for every x-value) must satisfy a given equation, then we can, under appropriate assumptions, solve the equations for one of the variables and treat the remaining variables as free variables. In practice the solution of an equation in closed form is usually not possible. For this reason we derive necessary conditions for solutions of variational problems with an equation as a side condition. We restrict ourselves here to the simplest case which may be stated as follows: Let the admissible functions be all pairs of smooth functions $y : [a, b] \mapsto \mathbb{R}$, $z : [a, b] \mapsto \mathbb{R}$ of the variable x which satisfy the boundary conditions

$$y(a) = y_a, \quad z(a) = z_a, \quad y(b) = y_b, \quad z(b) = z_b,$$

as well as the equation

$$g(x, y(x), z(x)) = 0 \quad \text{for all } x \in [a, b]. \tag{11.19}$$

Here g is a (partially) smooth function which is defined on an open set and has the property:

$$\begin{aligned}&\text{For each } (x, y, z) \text{ of the domain of definition of } g,\\ &\text{either } g_y(x, y, z) \neq 0 \text{ or } g_z(x, y, z) \neq 0.\end{aligned} \tag{11.20}$$

Among these function pairs (y, z) we seek those which give the variational integral

$$J(y, z) = \int_a^b f(x, y(x), z(x), y'(x), z'(x))\, dx$$

its smallest value. The equation $g(x,y,z)=0$ can be interpreted as the equation of a surface in space. Thus an admissible pair (y,z) represents a curve in space which lies on this surface. The variables y and z are no longer free but rather are bound by the equation. For a sensible problem, the relations $g(a,y_a,z_a)=0=g(b,y_b,z_b)$ must hold.

Just as is customary in the derivation of the necessary condition from the first variation, so we assume here that (y_0,z_0) is a two-fold smooth solution of the problem. That is, we consider a twice-continuously differentiable family $(\hat{y}(\cdot,\alpha),\hat{z}(\cdot,\alpha))$ (with parameter α) of admissible pairs such that for $\alpha=0$, $\hat{y}(x,0)=y_0(x)$ and $\hat{z}(x,0)=z_0(x)$ for all $x\in[a,b]$. Then

$$g(x,\hat{y}(x,\alpha),\hat{z}(x,\alpha))=0 \quad \text{for all } x \text{ and all } \alpha.$$

Partial differentiation with respect to α at the point $\alpha=0$ gives

$$g_y(x,y_0(x),z_0(x))\frac{\partial\hat{y}}{\partial\alpha}(x,0)+g_z(x,y_0(x),z_0(x))\frac{\partial\hat{z}}{\partial\alpha}(x,0)=0. \tag{11.21}$$

Here the usual procedure begins: The function Φ defined below has a (relative) minimum at the point $\alpha=0$

$$\begin{aligned}\Phi(\alpha)&=J(\hat{y}(\cdot,\alpha),\hat{z}(\cdot,\alpha)),\\ \Phi'(\alpha)=0&=\int_a^b\left\{\left[f_y^0(x)-\frac{d}{dx}f_{y'}^0(x)\right]\frac{\partial\hat{y}}{\partial\alpha}(x,0)\right.\\ &\qquad\left.+\left[f_z^0(x)-\frac{d}{dx}f_{z'}^0(x)\right]\frac{\partial\hat{z}}{\partial\alpha}(x,0)\right\}dx.\end{aligned} \tag{11.22}$$

As previously, we have used here the abbreviated forms

$$\begin{aligned}&f_y^0(x)=f_y(x,y_0(x),\ z_0(x),y_0'(x),\ z_0'(x)),\\ &\ldots,f_{z'}^0(x)=f_{z'}(x,y_0(x),\ z_0(x),y_0'(x),\ z_0'(x)).\end{aligned}$$

The functions $\frac{\partial\hat{y}}{\partial\alpha}(\cdot,0)$ and $\frac{\partial\hat{z}}{\partial\alpha}(\cdot,0)$ are linked to each other through the expression (11.21); thus we can proceed no further with the Fundamental Theorem of the Calculus of Variations. By virtue of assumption (11.20), however, we can solve (11.21) locally either for $\frac{\partial \hat{y}}{\partial \alpha}(x,0)$ or for $\frac{\partial \hat{z}}{\partial \alpha}(x,0)$. In order to simplify the calculations, we make assumption (11.20) a bit sharper and require

$$g_z^0(x) = g_z(x, y_0(x), z_0(x)) \neq 0 \quad \text{for all } x \in [a,b]. \tag{11.23}$$

Now (11.21) can be solved on the entire interval $[a,b]$ for $\frac{\partial \hat{z}}{\partial \alpha}(\cdot,0)$, and $\frac{\partial \hat{y}}{\partial \alpha}$ can be seen as a free variable. To find the solution here, we will employ a somewhat unconventional but, in this case, favourable method. Because of (11.23) there is a continuous function $\lambda : [a,b] \to \mathbb{R}$ such that for all $x \in [a,b]$

$$g_z^0(x)\lambda(x) = f_z^0(x) - \frac{d}{dx} f_{z'}^0(x). \tag{11.24}$$

If we multiply this equation with $\frac{\partial \hat{z}}{\partial x}(x,0)$ we obtain

$$\begin{aligned}\left[f_z^0(x) - \frac{d}{dx} f_x^0(x)\right] \frac{\partial \hat{z}}{\partial \alpha}(x,0) &= g_z^0(x)\frac{\partial \hat{z}}{\partial \alpha}(x,0)\lambda(x) \\ &= -\lambda(x) g_y^0(x) \frac{\partial \hat{y}}{\partial \alpha}(x,0).\end{aligned}$$

Then the substitution of this expression into the integral (11.22) gives

$$\Phi'(0) = 0 = \int_a^b \left[f_y^0(x) - \frac{d}{dx} f_{y'}^0(x) - \lambda(x) g_y^0(x)\right] \frac{\partial \hat{y}}{\partial \alpha}(x,0)\, dx.$$

But since $\frac{\partial \hat{y}}{\partial x}(x,0)$ may be regarded as an arbitrary smooth function with boundary values 0, as we noted above, the Fundamental Theorem of the Calculus of Variations yields the result

$$f_y^0(x) - \frac{d}{dx} f_{y'}^0(x) - \lambda(x) g_y^0(x) = 0. \tag{11.25}$$

The similarity of this equation with (11.24), which has a completely different meaning than (11.25), should be noted.

We can summarize the above results in the following way.

(11.26) Theorem
Let a variational problem with an equation $g(x, y, z) = 0$ as a side condition be given. Let the function g satisfy the assumption of (11.20) and let (y_0, z_0) be a two-fold smooth solution of the problem. Then there is a continuous function λ of x, called the Lagrange multiplier, such that (y_0, z_0) satisfies the Euler equations (11.24) and (11.25) of a variational problem with the integrand $f - \lambda g$.

Before we apply this theorem to two examples, we add an observation about the sharpened assumption (11.23). If (11.23) is disregarded while (11.20) holds, we proceed as follows: Clearly, for each $x \in [a, b]$ there is a sub interval $[x_0, x_1]$ with $x_0 < x < x_1$ in $[a, b]$ on which at least one of the functions $g_y(x, y_0(x), z_0(x))$ or $g_z(x, y_0(x), z_0(x))$ has no zero. The pair $(y_0, z_0) : [x_0, x_1] \mapsto \mathbb{R}^2$, however, is a solution of the variational problem

$$\int_{x_0}^{x_1} f dx \to \min,$$

$$\text{side condition } g(x, y, z) = 0,$$

with the boundary condition $y(x_0) = y_0(x_0)$, $z(x_0) = z_0(x_0)$, $y(x_1) = y_0(x_1)$, $z(x_1) = z_0(x_1)$.

Our proof is applicable to this problem and leads to the Euler differential equations for (y_0, z_0) at the point x, chosen arbitrarily from (a, b).

Example
We provide two solutions for the problem of the *motion of the mathematical pendulum.*

1. As a variational problem without side conditions: If we let $\phi(t)$ be the angle which the string forms with the positive y-axis, then the position $(x(t), y(t))$ of the point-mass can be expressed as

$$x(t) = L \sin \phi(t), \quad y = L - L \cos \phi(t).$$

Then the action integral is

$$J(\phi) = \int_{t_0}^{t_1} [T(t) - U(t)]\, dt$$
$$= \int_{t_0}^{t_1} [\tfrac{1}{2}L^2\phi'^2(t) + gL(1 - \cos\phi(t))]dt.$$

The boundary conditions for ϕ can be obtained from the boundary conditions for x and y. The function ϕ thus satisfies the Euler differential equation of this variational problem, namely

$$L\phi''(t) + g\sin\phi(t) = 0. \qquad (11.27)$$

As long as only very small angles are involved, an approximate solution of (11.27) can be obtained from the differential equation

$$L\phi''(t) + g\phi(t) = 0.$$

2. As a variational problem with side condition: In this case the variational integral is

$$J(x, y) = m\int_{t_0}^{t_1} [\tfrac{1}{2}(x'^2(t) + y'^2(t)) + gy(t)]\, dt$$

while the side condition is $x^2(t) + (y(t) - L)^2 - L^2 = 0$. The assumptions of Theorem (11.26) are satisfied. Thus there is a continuous function λ such that the actual motion satisfies the equations

$$x''(t) + 2\lambda(t)x(t) = 0,$$
$$y''(t) - g + 2\lambda(t)(y(t) - L) = 0,$$

and the side condition

$$x^2(t) + (y(t) - L)^2 - L^2 = 0.$$

Example

Find the shortest curve from A to B on the surface $g(x, y, z) = 0$.

Let the surface $\mathbb{F}$ be given by its equation $g(x, y, z) = 0$ and suppose the assumptions (11.20) are satisfied; i.e. the surface has a tangent plane at each point.

Among all the curves k on the surface $\mathbb{F}$ which can be expressed with x as a parameter, find those which join the points $A = (a, y_a, z_a)$ and $B = (b, y_b, z_b), a < b$, on the surface $\mathbb{F}$ and which have the smallest length. Equivalently, we seek curves such that

$$J(y, z) = \int_a^b \sqrt{1 + y'^2(x) + z'^2(x)}\, dx \to \min$$

under the side condition $g(x, y(x), z(x)) = 0$. Theorem (11.26) gives us the following necessary condition for solutions: There is a continuous function λ of the variable x such that

$$\frac{d}{dx}\frac{y'}{\sqrt{1 + y'^2 + z'^2}}(x) - \lambda(x) g_y(x, y(x), z(x)) = 0,$$

$$\frac{d}{dx}\frac{z'}{\sqrt{1 + y'^2 + z'^2}}(x) - \lambda(x) g_z(x, y(x), z(x)) = 0.$$

11.4 LAGRANGE'S PROBLEM

By *'Lagrange's problem'* we mean a variational problem which has a differential equation as a side condition. We will note the multiplier rule here and then apply it to an example. For the proof of this not entirely simple rule, we refer the reader to the literature, e.g. [3].

The Lagrange problem may be formulated as follows: Consider all smooth vector functions

$$\underline{y} = (y_i)_{i=1,2,\cdots,n} : [a, b] \mapsto \mathbb{R}^n$$

which satisfy the admissibility conditions consisting of:

(a) the boundary conditions which, for example, for the fixed boundary problem, require that

$$y_i(a) = y_{ia}, \quad y_i(b) = y_{ib}, \quad i = 1, 2, \ldots, n; \tag{11.28}$$

and

(b) the r differential equations

$$g_\alpha(x, y_i(x), y_i'(x)) = 0, \quad \alpha = 1, 2, \ldots, r, \tag{11.29}$$

where the functions g_α are defined on an open set and are three-fold (partially) smooth.

Among these functions find those which give the variational integral

$$J(\underline{y}) = \int_a^b f(x, y_i(x), y_i'(x))\, dx$$

its smallest value.

(11.30) Theorem (Multiplier rule)
Let $y^0 : [a, b] \mapsto \mathbb{R}^n$ be a two-fold smooth solution of the problem of Lagrange with the differential equations $g_\alpha(x, y_i, y_i') = 0$, $\alpha = 1, \ldots, r$, as a side condition. Let these differential equations be independent along $\underline{y}^0$; i.e. for all $x \in [a, b]$, the rank of the matrix

$$\left(\frac{\partial g_\alpha}{\partial y_i'}(x, \underline{y}^0(x), \underline{y}^{0\prime}(x))\right), \quad \begin{matrix} \alpha = 1, \ldots, r, \\ i = 1, \ldots, n, \end{matrix} \tag{11.31}$$

shall be equal to r. Then there are $r + 1$ smooth functions, the Lagrange multipliers, $\lambda_0, \lambda_1, \ldots, \lambda_r$, which have no common zeros in $[a, b]$ and are such that $\underline{y}_0$ satisfies the Euler differential equations of a variational problem without side conditions and with integrand

$$\begin{aligned} f^*(x, y_i, y_i') = {} & \lambda_0(x) f(x, y_i, y_i') \\ & + \lambda_1(x) g_1(x, y_i, y_i') + \cdots \\ & + \lambda_r(x) g_r(x, y_i, y_i'). \end{aligned} \tag{11.32}$$

The multiplier λ_0 is a constant.

We will use the example of the equilibrium state of the homogeneous chain to show the way in which this theorem works. The mathematical model for the problem was derived in (11.4). We seek functions x and y of the variable s such that

$$J(x,y) = g\int_0^L y(s)\,ds \to \min \quad (g = \text{ constant})$$

under the boundary conditions

$$x(0) = a, \quad y(0) = y_a, \quad x(L) = b, \quad y(L) = y_b, \quad a < b,$$

and the side condition

$$x'^2(s) + y'^2(s) = 1 \quad \text{for all } s \in [0, L]. \tag{11.33}$$

The integrand f^* of the multiplier rule is then

$$f^*(s,x,y,x',y') = \lambda_0 gy + \lambda_1(s)(x'^2 + y'^2 - 1).$$

If there is, therefore, a two-fold smooth solution x^0, y^0 of this Lagrange problem, then the functions x^0 and y^0 must satisfy the following differential equations in which λ_0 is an unknown constant and λ_1 is an unknown function

$$\begin{aligned} \lambda_0 g - \frac{d}{ds}2(\lambda_1 y') &= 0,\\ -\frac{d}{ds}2(\lambda_1 x') &= 0. \end{aligned}$$

We can integrate these differential equations to obtain

$$\begin{aligned} \lambda_1(s)x'(s) &= \text{constant} = c_1,\\ \lambda_1(s)y'(s) &= \tfrac{1}{2}\lambda_0 gs + c, \end{aligned} \tag{11.34}$$

where c_1 and c are constants of integration.

We can exclude the case $c_1 = 0$. Since the multipliers λ_0 and λ_1 are determined uniquely only up to a common factor (cf.

Theorem (11.30)), we can set $c_1 = 1$. We now bring the side conditions (11.35) into the picture

$$\lambda_1^2(s)[x'^2(s) + y'^2(s)] = \lambda_1^2(s) = 1 + (\tfrac{1}{2}\lambda_0 g s + c)^2. \qquad (11.35)$$

Thus λ_1 is determined except for sign. If we substitute

$$\lambda_1(s) = \pm\sqrt{a^2(s+s_0)^2+1}, \; a = \tfrac{1}{2}\lambda_0 g, \quad s_0 = \frac{c}{a},$$

into (11.34), $\lambda_0 \neq 0$, we obtain

$$\begin{aligned} x(s) &= \pm\frac{1}{a} \operatorname{arg\,sinh} a(s+s_0), \\ y(s) &= \pm\frac{1}{a}\sqrt{a^2(s+s_0)^2+1} \end{aligned}$$

or, since $s + s_0 = \frac{1}{a}\sinh ax(s)$, $y(s) = \frac{1}{a}\cosh ax(s)$.

These equations provide a parametric representation of the catenary. From (11.35) it follows that if $\lambda_0 = 0$, then λ_1 is a constant. Thus because of (11.34) x and y must be linear functions of s and so represent a line segment. The constants λ_0 and c as well as a and s_0 must be deduced from the boundary conditions.

12

Introduction to the direct methods of the calculus of variations

---

Only in relatively few cases is the Euler differential equation easy to solve. If an exact solution to the Euler differential equation with the prescribed boundary conditions cannot be found, then we must appeal to approximation methods to solve the variational problem. There are several procedures available – all classified under the general heading '*direct methods* of the calculus of variations'. We will show one of these procedures (Ritz's) here and will illustrate its use with an example. For a systematic treatment of this direct method, the reader is referred to the literature, e.g. [15].

Direct methods have also found use in the solution of boundary value problems which, at first glance, are independent of any variational problem. In such cases we look for a variational problem whose Euler equation coincides with the ordinary or partial differential equation of the boundary value problem. It should also be noted that direct methods are of significant theoretical interest, since they provide existence proofs for solutions of variational problems.

We turn now to a description of *Ritz's method.* Let a variational problem $J \to \min$ be given without side conditions. That is, consider simply a fixed boundary problem with only one unknown function. The procedure is at the outset the same for simple as it is for multiple integrals. Thus, in this instance, we do not need to commit ourselves. We select a sequence $z_n, n = 0, 1, 2, \ldots,$ of functions with the following three properties:

(i) z_0 is any admissible function which, in particular, satisfies the prescribed boundary conditions.

(ii) The functions z_n have the value zero on the boundary of the domain of integration and are (possibly piecewise) smooth.

Then every function y_n of the form

$$y_n = z_0 + c_1 z_1 + c_2 z_2 + \cdots + c_n z_n \tag{12.1}$$

is an admissible function of the variational problem; here the c_i are arbitrary constants and n is an arbitrary natural number.

(iii) For every admissible function y there can be found a function y_n of the form (12.1) such that the difference $J(y) - J(y_n)$ is arbitrarily small. The value $|J(y) - J(y_n)|$ will be used as a measure of the closeness of the approximation.

Under the assumption which we have placed on the integrand f of a variational integral, the value $|J(y) - J(y_n)|$ is small if the distance $d_1(y, y_n)$ from y to y_n is sufficiently small. The converse of this statement does not hold, for there are instances for which $|J(y_n) - J(y)| \to 0$ as $n \to \infty$ even though $d_1(y, y_n)$ does not converge to zero.

We assume now that $J(y)$ is bounded from below. Then there exists a sequence of admissible functions $\tilde{y}_n$ such that as $n \to \infty$

$$J(\tilde{y}_n) \to J_0 = \inf J(y) \geq K > -\infty,$$

where the infimum is to be taken over all admissible functions. Such a sequence $(\tilde{y}_n)$ is called a *minimizing sequence* and the number J_0 is then the minimal value of the variational integral.

For the functions $\tilde{y}_n$, Ritz's method employs admissible functions of the form (12.1). Because the calculations are quite extensive, it is prudent to give attention to the functions z_n of the problem so that the convergence will be as fast as possible. In order to adjudge the closeness of the approximation, we need an estimate of the error $J(\tilde{y}_n) - J_0$. Since J_0 is in general not known, we must rely on finding a good lower bound for J_0. It is with this problem that we will be preoccupied in the example below. We have found a satisfactory solution to the variational problem of the boundary value problem only if the following questions are answered: Does the minimizing sequence (y_n) converge (uniformly?) to a limit function y^0? If it does not, does a subsequence of the sequence $(\tilde{y}_n)$ at least converge? Is the limit function y_0 an admissible function? That is, is it – or possibly its derivative – (piecewise) smooth, and is it a solution of the Euler differential equation of the variational problem? It is possible to answer these questions for large classes of function spaces with the methods of functional analysis.

As an example, we will consider the variational problem

$$\begin{aligned} J(y) = \int_a^b [&d_1(x)y'^2(x) + 2d_2(x)y'(x)y(x) \\ &+ d_3(x)y^2(x)]\,dx \to \min, \\ &y(a) = y_a, \quad y(b) = y_b, \end{aligned} \tag{12.2}$$

where d_1, d_2 and d_3 are arbitrary smooth functions which for $x \in [a, b]$ satisfy the following conditions:

$$d_1(x) > 0, \quad d_1(x)d_3(x) - d_2^2(x) > 0. \tag{12.3}$$

Because of (12.3), the integrand of the problem is convex in (y, y'). The Euler differential equation for (12.2) is a linear, homogeneous differential equation of second order

$$d_1(x)y''(x) + d_1'(x)y'(x) + (d_2'(x) - d_3(x))y(x) = 0,$$

which may be put in the form

$$y''(x) + A(x)y'(x) + B(x)y(x) = 0, \tag{12.4}$$

an equation with continuous coefficient functions A and B. Hence it is possible to determine d_1, d_2 and d_3 so that (12.4) is the Euler differential equation for (12.2). In any event, the integrand so determined will not in every case be convex. There are many possibilities for the choice of the functions z_n with properties (i), (ii) and (iii). So, for example,

$$z_0(x) = y_a + (x-a)\frac{y_b - y_a}{b-a} \quad \text{for} \quad x \in [a,b],$$

$$z_n(x) = \sin\frac{n\pi(x-a)}{b-a} \quad \text{for} \quad x \in [a,b], n = 1, 2, \ldots.$$

We see immediately that the sequence (z_n) satisfies properties (i) and (ii). We will show in (12.5) that it also satisfies (iii). But now we want to determine a minimizing sequence

$$\tilde{y}_n = z_0 + c_{1n}z_1 + c_{2n}z_2 + \cdots + c_{nn}z_n, \qquad n = 1, 2, \ldots.$$

To do this, substitute a function y_n of the form $y_n = z_0 + c_1z_1 + c_2z_2 + \cdots + c_nz_n$ in the variational integral. $J(y_n)$ now depends only on the coefficients $c_1, c_2, \ldots, c_n$, and so is a function of n variables. It is seen that $J(y_n)$ is a quadratic polynomial in the c_i's. In order to find the minimum of this polynomial, we set each of its partial derivatives with respect to the c_i equal to zero. We now have a system of n linear non-homogeneous equations for the n unknowns $c_1, c_2, \ldots, c_n$. Because of the convexity assumption for the integrand, the determinant of the coefficient matrix of this system is different from zero. Thus, the system has a unique solution $c_{1n}, c_{2n}, c_{3n}, \ldots, c_{nn}$. From this it follows that among all the functions of the form $z_0 + c_1z_1 + \cdots + c_nz_n$, the function $\tilde{y}_n = z_0 + c_{1n}z_1 + c_{2n}z_2 + \cdots + c_{nn}z_n$ yields the smallest value for the variational integral J.

For this class of variational problems, the minimizing sequence $(\tilde{y}_n)$ converges uniformly to a two-fold smooth function y^0 which satisfies the Euler differential equation (cf. [14]).

Exercise
Find the first functions $\tilde{y}_n$ of a minimizing sequence for the variational problem

$$J(y) = \int_{-\pi}^{\pi} [y'^2(x) + y^2(x)]\, dx \longrightarrow \min,$$
$$y(-\pi) = -1, \quad y(\pi) = 1.$$

Find also the exact solution and compare the results.

(12.5) A Supplement

Because of assumption (12.3), there is no solution with corners. Thus we can look for the solution among smooth admissible functions y which, as we will now show, satisfy the approximation property (iii). We continue the function $y - z_0$ to the interval $[a-(b-a), a]$ by means of the formula

$$y(a-x) - z_0(a-x) = -[y(a+x) - z_0(a+x)]$$
$$\text{for } 0 \leq x \leq b-a,$$

and thereafter to all of $\mathbb{R}$ as a periodic function with period $2(b-a)$. The function $y - z$ has zeros at a and b and is smooth on $\mathbb{R}$. Thus, $y - z_0$ may be developed as a uniformly convergent Fourier series

$$y(x) - z_0(x) = \sum_{i-1}^{\infty} c_i \sin \frac{i\pi(x-a)}{b-a}.$$

(Notice that $y - z_0$ is an odd function of $(x-a)$.) For the nth approximation y_n of y we can thus choose y_n to be z_0 plus the n th partial sum of the Fourier series. Clearly, $J(y) - J(y_n) \rightarrow 0$ as $n \longrightarrow \infty$.

In order to evaluate the closeness of the approximation and to estimate the error, we will derive here lower limits for the minimum value J_0 of the variational problem $J(y) \rightarrow \min$. To this end, we associate with the variational problem $J(y) \rightarrow \min$ a new variational problem $J_D(p) \rightarrow \max$, which has the two following properties.

For arbitrary admissible functions y and p,

$$J_D(p) \le J(y) \tag{12.6}$$

and

$$\sup J_D(p) = J_0 = \inf J(y). \tag{12.7}$$

The variational problem $J_D(p) \to \max$ with (12.6) and (12.7) is called the *dual problem* of $J(y) \to \min$. Two variational problems $J(y) \to \min$ and $J_D(p) \to \max$ which are related by (12.6) and (12.7) are called *complementary variational problems.* The values $J_D(p)$ for arbitrary admissible functions p of the dual problem are lower bounds for J_0 because of (12.6). The error $|J(y) - J_0|$ can be approximated by

$$|J(y) - J_0| \le |J(y) - J_D(p)|.$$

By means of direct methods one can, furthermore, determine admissible functions p_n of the dual problem such that $J_D(p_n)$ converges to J_0, thereby improving the lower bounds. In what follows, we will construct a dual problem for a fixed boundary problem with a simple integral. We will assume that the integrand f is convex with respect to y and y' and that for all x the equations

$$p = f_{y'}(x, y, y'), \quad q = f_y(x, y, y'), \tag{12.8}$$

can be solved uniquely for y and y'; i.e.

$$y = y^*(x, p, q), \quad y' = z^*(x, p, q), \tag{12.9}$$

where the functions y^* and z^* are smooth (partially) with respect to x, p and q. The quantities p and q are new variables. To avoid misunderstandings, we will write z instead of y' but retain the notation $f_{y'}$ to denote the partial derivative of f with respect to the third variable of f.

Equation (12.8) defines an invertible transformation of xyz-space into the xpq-space. The integrand f_D of the dual problem is given by

$$\begin{aligned} f_D(x, p, q) =& f(x, y^*(x, p, q), z^*(x, p, q)) \\ & - pz^*(x, p, q) - qy^*(x, p, q), \end{aligned} \tag{12.10}$$

or, more concisely, $f_D(x,p,q) = f(x,y,z) - pz - qy$, where (x,y,z) and (x,p,q) are related to each other through (12.8) and (12.9).

(12.11) Theorem

Let there be given a fixed boundary problem $J(y) \to \min$ *with the boundary conditions* $y(a) = y_a$ *and* $y(b) = y_b$ *which satisfy the assumptions stated above. Also, let* y_0 *be an extremal on* $[a,b]$ *which satisfies the boundary conditions. Then the Lagrange problem,*

$$J_D(p) = \int_a^b f_D(x, p(x), p'(x))\, dx + y_b p(b) - y_a p(a) \to \max, \tag{12.12}$$

whose admissible functions $p(x)$ *satisfy the two following differential equations* (12.13) *and* (12.14), *is a dual problem for* $J(y) \to \min$

$$\frac{d}{dx} y^*(\cdot, p, p')(x) = z^*(x, p(x), p'(x)), \tag{12.13}$$

$$\frac{d}{dx} p(x) = p'(x) = f_y(x, y^*(x, p(x), p'(x)), z^*(x, p(x), p'(x))). \tag{12.14}$$

Proof

Since the integrand f of the problem $J(y) \to \min$ is convex in (y, y'), it follows for all values of $x, y, z, \tilde{y}$ and $\tilde{z}$ that

$$\begin{aligned} f(x, \tilde{y}, \tilde{z}) \geq\ & f(x,y,z) + f_y(x,y,z)(\tilde{y} - y) \\ & + f_{y'}(x,y,z)(\tilde{z} - z). \end{aligned} \tag{12.15}$$

Let $\tilde{y}$ be an arbitrary admissible function for the problem $J(y) \to \min$ and let p be an arbitrary admissible function for the Lagrange problem of the theorem. For p, define the function y by means of

$$y(x) = y^*(x, p(x), p'(x)).$$

Then, because of (12.14), the function y is an extremal of the problem $J(y) \to \min$. The triples $(x, p(x), p'(x))$ and $(x, y(x), y'(x))$ are associated with each other through (12.8) and (12.9). From the inequality (12.15) we obtain

$$\begin{aligned} f(x,\tilde{y}(x),\tilde{y}'(x)) \geq\ & f(x,y(x),y'(x)) \\ & + f_y(x,y(x),y'(x))(\tilde{y}(x)-y(x)) \\ & + f_{y'}(x,y(x),y'(x))(\tilde{y}'(x)-y'(x)) \\ =\ & f_D(x,p(x),p'(x)) + \tilde{y}(x)p'(x) + \tilde{y}'(x)p(x). \end{aligned}$$

Integration from a to b yields

$$\begin{aligned} J(\tilde{y}) &= \int_a^b f(x,\tilde{y}(x),\tilde{y}'(x))\,dx \\ &\geq \int_a^b f_D(x,p(x),p'(x))\,dx + [\tilde{y}(x)p(x)]_a^b \\ &= J_D(p). \end{aligned}$$

Since this inequality holds for arbitrary admissible functions $\tilde{y}$ and p, property (12.6) follows.

Because f is convex with respect to the variables y and $z = y'$, the extremal y_0 is a solution for the problem $J(y) \to \min$ (cf. Theorem (3.16)). With the extremal y_0 we associate the function p_0 by means of the equation

$$p_0(x) = f_{y'}(x, y_0(x), y_0'(x)).$$

Here p_0 is an admissible function of the Lagrange problem. Then it follows that

$$\begin{aligned} J_0 &= \inf J(y) = J(y_0) \\ &= \int_a^b f(x,y_0(x),y_0'(x))\,dx \\ &= \int_a^b [f_D(x,p_0(x),p_0'(x)) + p_0(x)y_0'(x) + p_0'(x)y_0(x)]\,dx \\ &= \int_a^b f_D(x,p_0(x),p_0'(x))\,dx + p_0(b)y_b - p_0(a)y_a \\ &= J_D(p_0). \end{aligned}$$

Thus (12.7) is also shown and we may affirm that the Lagrange problem given in the theorem is a dual of the problem $J(y) \to \min$.

We can obtain a dual problem for variational problems with several unknown functions or with multiple integrals in a similar way, as long as the integrand is convex. The transformation (12.8) needs to be generalized in a sensible way.

We return again to the example of (12.2)

$$\begin{aligned} J(y) = \int_a^b [d_1(x)y'^2(x) + 2d_2(x)y(x)y'(x) \\ + d_3(x)y^2(x)]\,dx \to \min, \\ y(a) = y_a, \quad y(b) = y_b. \end{aligned} \tag{12.16}$$

We impose on the coefficients the same requirements as those given above; (12.3) especially must be satisfied. The integrand is then convex in y and y' and the transformation (12.8)

$$p = 2d_1(x)z + 2d_2(x)y, \quad q = 2d_2(x)z + 2d_3(x)y,$$

is invertible

$$\begin{aligned} y = y^*(x,p,q) &= \frac{1}{2D(x)}(-d_2(x)p + d_1(x)q), \\ z = z^*(x,p,q) &= \frac{1}{2D(x)}(d_3(x)p - d_2(x)q), \end{aligned}$$

where $D(x) = d_1(x)d_3(x) - d_2^2(x) > 0$.

The integrand of the dual is now

$$f_D(x,p,q) = -\frac{1}{4D(x)}[d_3(x)p^2 - 2d_2(x)pq + d_1(x)q^2].$$

The side condition (12.13) is satisfied identically; then from (12.12) we infer

$$d_1 p'' + \left(\frac{d_1}{D}\right)' Dp' - \left[d_3 + \left(\frac{d_3}{D}\right)' D\right] p = 0. \tag{12.17}$$

This linear homogeneous differential equation of second order is, in general, equally as difficult or easy to solve as the Euler equation of the initial problem. But even then, if one cannot determine exactly the admissible functions p of the dual problem, one has the possibility of calculating a lower bound for the minimum value J_0 of the variational integral J by using an approximating solution p_n of (12.17).

Exercise

For the variational problem

$$J(y) = \int_{-\pi}^{\pi} [y'^2(x)+y^2(x)]\, dx \to \min, \quad y(-\pi) = -1, \quad y(\pi) = 1,$$

find a lower bound for the minimal value J_0 of the variational problem by using the idea of the dual problem.

Bibliography

1. *Abhandlung über Variations-Rechnung*, first part: *Abhandlungen von Joh. Bernoulli* (1696), Jac. Bernoulli (1697) and Leonhard Euler (1744), second part: *Abhandlungen von Lagrange* (1762, 1770), Legendre (1786) und Jacobi (1837), published by P. Stäckel in the series *Ostwald's Klassiker der Exakten Wissenschaften* Vols. 46 and 47, Leipzig.

2. Bliss, G.A., *Lectures on the Calculus of Variations*, University of Chicago Press, 1946.

3. Bolza, O., *Lectures on the Calculus of Variations*, reprinted by G.E. Stechert and Co., New York, 1931.

4. Carathéodory, C., *Variationsrechnung und partielle Differentialgleichungen erster Ordnung*, Teubner, Berlin, 1935.

5. Courant, R. and D. Hilbert, *Methods of Mathematical Physics*, Vol. I, Interscience Publishers, Inc., New York, 1953, Chapters 4 and 6.

6. Funk, P., *Variationsrechnung und ihre Anwendungen in Physik und Technik*, Springer, Berlin, 1968 (Third Edition).

7. Grüss, G., *Variationsrechnung*, Quelle und Meyer, Heidelberg, 1955 (Second Edition).

8. Hadley, G. and M.C. Kemp, *Variational Methods in Economics*, North Holland Publishing Co., Amsterdam, 1971.

9. Klingbiel, E., *Variationsrechnung*, Bibliographisches Institut, Mannheim, 1977.

10. Klötzler, R., *Mehrdimensionale Variationsrechnung*, Birkhäuser Verlag, Basil, 1970.

11. Mikhlin, S.G., *Variationsmethoden der Mathematischen Physik*, Akademie-Verlag, Berlin, 1962.

12. Morrey, C.B., *Multiple Integrals in the Calculus of Variations*, Die Grundlehren der mathematischen Wissenschaften, Vol. 30, Springer, Berlin, 1966.

13. Morse, M., *The Calculus of Variations in the Large*, A.M.S. Colloquium Publ. No. 18, Providence, R.I., 1934.

14. Ritz, W., Über eine neue Methode zur Lösung gewisser Variationsprobleme der mathematischen Physik, *Journal für die reine und angewandte Mathematik*, Vol. 135 (1909), pp. 1-61.

15. Velte, W., *Direkte Methoden der Variationsrechnung*, Teubner, Stuttgart, 1976.

Index